climate policy

■ **The leading international, peer-reviewed journal on responses to climate change**

VOLUME 8 SUPPLEMENT 2008

climate policy

modelling long-term scenarios for low-carbon societies

First published 2008
by Earthscan

Published 2014 by Routledge
2 Park Square, Milton Park, Abingdon, Oxon OX14 4RN
711 Third Avenue, New York, NY, 10017, USA

*Routledge is an imprint of the Taylor & Francis Group,
an informa business*

Climate Policy is editorially independent. The editorial
administration of the journal is supported by Climate
Strategies (a not-for-profit research network), the
French region Ile de France and Le Centre National de
la Recherche Scientifique (CNRS) in France.

Climate Policy is indexed in Thomson ISI Social Sciences
Citation Index.

Abstracting services which cover this title include:
Elsevier Scopus, Geobase, International Political
Science Abstracts and Econlit.

The Publishers acknowledge the generous support of
Shell Foundation and Climate Strategies for the
publication of this journal.

ISBN 13: 978-1-844-07594-2 (hbk)

CENTRE NATIONAL
DE LA RECHERCHE
SCIENTIFIQUE

Climate
Strategies

PREFACE
S3–S4 **Low-Carbon Society (LCS) modelling**
NEIL STRACHAN, TIM FOXON, JUNICHI FUJINO

EDITORIAL
S5–S16 **Policies and practices for a low-carbon society**
JIM SKEA, SHUZO NISHIOKA

SYNTHESIS
S17–S29 **Policy implications from the Low-Carbon Society (LCS)
modelling project**
NEIL STRACHAN, TIM FOXON, JUNICHI FUJINO

RESEARCH
S30–S45 **Achieving the G8 50% target: modelling induced
and accelerated technological change using the
macro-econometric model E3MG**
TERRY BARKER, S. SERBAN SCRIECIU, TIM FOXON

S46–S59 **Global emission reductions through a sectoral intensity
target scheme**
KEIGO AKIMOTO, FUMINORI SANO, JUNICHIRO ODA, TAKASHI
HOMMA, ULLASH KUMAR ROUT, TOSHIMASA TOMODA

S60–S75 **A global perspective to achieve a low-carbon society
(LCS): scenario analysis with the ETSAP-TIAM model**
UWE REMME, MARKUS BLESL

S76–S92 **Implications for the USA of stabilization of radiative
forcing at 3.4 W/m²**
JAE EDMONDS, LEON CLARKE, MARSHALL WISE, HUGH
PITCHER, STEVE SMITH

S93–S107 **Permit sellers, permit buyers: China and Canada's roles
in a global low-carbon society**
CHRIS BATAILLE, JIANJUN TU, MARK JACCARD

S108–S124 **Back-casting analysis for 70% emission reduction in
Japan by 2050**
JUNICHI FUJINO, GO HIBINO, TOMOKI EHARA, YUZURU
MATSUOKA, TOSHIHIKO MASUI, MIKIKO KAINUMA

S125–S139 **The role of international drivers on UK scenarios of a
low-carbon society**
NEIL STRACHAN, STEPHEN PYE, NICHOLAS HUGHES

S140–S155 **Effects of carbon tax on greenhouse gas mitigation in
Thailand**
RAM M. SHRESTHA, SHREEKAR PRADHAN, MIGARA H.
LIYANAGE

S156–S176 **Low-carbon society scenarios for India**
P.R. SHUKLA, SUBASH DHAR, DIPTIRANJAN MAHAPATRA

Aims and scope

Climate Policy presents the highest quality refereed research and analysis on the policy issues raised by climate change, and provides a forum for commentary and debate. It addresses both the mitigation of, and adaptation to, climate change, within and between the different regions of the world. It encourages a trans-disciplinary approach to these issues at international, regional, national and sectoral levels.

The journal aims to make complex, policy-related analysis of climate change issues accessible to a wide audience, including those actors involved in:

- research and the commissioning of policy-relevant research
- policy and strategy formulation/implementation by local and national governments;
- the interactions and impacts of climate policies and strategies on business and society, and their responses, in different nations and sectors;
- international negotiations including, but not limited to, the UN Framework Convention on Climate Change, the Kyoto Protocol, other processes.

Climate Policy thus aims to build on its academic base so as to inject new insights and facilitate informed debate within and between, these diverse constituencies.

Types of contribution

Climate Policy publishes a variety of contributions:

Peer reviewed articles

Peer reviewed articles present academic, evidence-based research on climate policy issues:

- **Research articles** (4–6000 words) present original high quality research
- **Synthesis articles** (6–8000 words) present a survey and syntheses of the state of knowledge and key issues in a particular area of relevance to climate policy, including scientific, economic, environmental, institutional, political, social or ethical issues.
- **Policy analysis articles** (1–3000 words) present evidence-based objective analysis of policy that is embedded within an existing literature and context.

Research and synthesis articles are subject to rigorous double-blind multiple academic peer review; policy analysis articles are also fully peer-reviewed.

Outlook

The Outlook section presents timely, relevant analysis and commentary, for a wide climate policy community, and includes:

- **Perspectives** from senior decisionmakers
- **Insights** from independent commentators on policy processes, positions, options and debates
- **Records** of important new agreements, legislation and other developments including analysis of key events
- **Feedback** on earlier material published in *Climate Policy*

Climate Policy Outlook contains both commissioned and submitted papers, subject to editorial and light external review, generally in the range 500-2500 words though longer pieces may be considered on an exceptional basis.

Climate Policy also carries **country studies** and **book reviews**, and publishes **special thematic issues** on particular topics.

Topics covered

Topics covered by *Climate Policy* include (but are not limited to):

- Analysis of mitigation or adaptation policies and strategies (at macro-, meso- and/or micro- scales)
- Studies of implementation and prospects in different countries and industrial sectors
- Sectoral options and strategies for meeting policy targets
- Studies on regional differences including North-South issues
- Policy and economic aspects of intergenerational and intragenerational equity
- Applications of integrated assessment to specific policy issues
- Policy and quantitative aspects of land-use and forestry
- Design of the Kyoto mechanisms and their implications
- Analysis of corporate strategies for climate change
- Socio-political analysis of prospects for the UNFCCC system
- Economic and political aspects of developing country policy formation, action and involvement
- Social studies of climate change, including public perception, where policy implications are derived
- Local resilience, adaptation and insurance measures: extreme events and gradual change
- National and international adaptation and coping with impacts, including migration, natural resource allocation and use, etc.
- Policy formulation processes, including negotiation, public consultation, political processes and 'bottom-up' approaches

Authors' charter

This journal is committed to maintain the highest editorial standards and continuous improvement. As part of moving to an online submission system, we are implementing an Authors' Charter (see www.climatepolicy.com) to make our procedures and policies explicit and to help authors understand the editorial process and what to expect. As part of maintaining the highest standards, *Climate Policy* asks all authors, editors and reviewers to disclose any relationship (e.g. financial, economic or institutional) that could be perceived as affecting the integrity of the scientific process.

climate
policy

■ preface

Low-Carbon Society (LCS) modelling

NEIL STRACHAN[1]*, TIM FOXON[2], JUNICHI FUJINO[3]

[1] Department of Geography, King's College London, UK
[2] Sustainability Research Institute, School of Earth and Environment, University of Leeds, UK
[3] National Institute for Environmental Studies, Japan

What are plausible visions of a low-carbon society, what options exist to achieve the transition to a low-carbon society, and what are the implications of those different options? This *Climate Policy* supplement reports the research findings and policy implications from an international comparative exercise to model scenarios of long-term low-carbon societies. This was undertaken within the framework and with the support of the Japan–UK research project 'Low-Carbon Society (LCS) Scenarios Towards 2050'. This project was jointly promoted by the Ministry of Environment Japan (MoEJ) and the Department for Environment, Food and Rural Affairs in the UK (DEFRA) with the aim of informing the Gleneagles Dialogue on Climate Change, Clean Energy and Sustainable Development, which was established during the UK's 2005 presidency of the G8, with progress to be reported and discussed during Japan's presidency in 2008. Following the first LCS workshop in Tokyo, it was agreed to bring together an international group of climate–economy modellers, including representatives from developing countries, to a meeting at the UK Energy Research Centre in December 2006 to plan a model comparison exercise. The results of that exercise, involving researchers from the UK, Japan, Germany, the USA, Canada, Thailand and India, are reported in this supplement. There are two introductory papers – first, the LCS workshop co-chairs, Jim Skea and Shuzo Nishioka, discuss the broader policy context of the LCS project, and second, we discuss the common themes for policymakers arising from the results.

The LCS modelling exercise described here takes its lead from the declaration issued during the 2007 G8 summit in Heiligendamm supporting a global target of a 50% reduction in GHGs by 2050. This corresponds to the more stringent stabilization pathways envisioned under the IPCC Fourth Assessment Report, but which, so far, have been subject to relatively little scenario analysis. The exercise used a range of global and national energy models; macroeconomic, technology-focused and hybrid approaches. Each modelling team investigated at least three scenarios: a *Base case*, a *Carbon price* case (rising to $100/tCO$_2$ by 2050), and one or more *Carbon-plus* cases to analyse what additional measures may be needed to achieve a LCS scenario consistent with a 50% reduction in global CO$_2$ emissions by 2050. Individual modelling assessments focused on independently chosen core drivers and utilized the models' particular strengths.

We believe that the results of this exercise will be valuable to national and international policy makers and can usefully inform the discussions on the Gleneagles Dialogue during Japan's G8 presidency. The papers address the levels of technological progress and complementary behavioural change needed to achieve a low-carbon society; issues relating to timing of actions and the role of emissions targets; the economic costs and benefits of different pathways to a low-carbon society;

■ *Corresponding author. *E-mail*: neil.strachan@kcl.ac.uk

doi:10.3763/cpol.2008.0538 © 2008 Earthscan ISSN: 1469-3062 (print), 1752-7457 (online) www.climatepolicy.com

the particular challenges facing developing countries to achieve LCS in light of their projected economic growth and energy use requirements; and the consequent need for international cooperation, notably in flexible burden sharing under international emissions trading regimes.

Many have contributed to the success of this research project, including the modelling teams, the editors and referees of *Climate Policy*, and the UKERC Meeting Place for organizing the LCS meeting within the Annual Energy Modelling Conference in Oxford, UK, in December 2006. In addition, we are particularly grateful for funding of this special issue from the UK Energy Research Centre (UKERC), the UK Department of Environment, Food and Rural Affairs (DEFRA), and the Japanese National Institute for Environmental Studies (NIES).

climate
policy

■ editorial

Policies and practices for a low-carbon society

JIM SKEA[1]*, SHUZO NISHIOKA[2]

[1] UK Energy Research Centre, 58 Prince's Gate, London SW7 2PG, UK
[2] National Institute for Environmental Studies, 16-2 Onogawa, Tsukuba-City, Ibaraki, 305-8506, Japan

Introduction

In February 2006, the Ministry of Environment (MOE) Japan and the Department of Environment, Food and Rural Affairs (DEFRA) in the UK set in motion an ambitious research project aimed at informing the Gleneagles Dialogue on Climate Change, Clean Energy and Sustainable Development established during the UK's 2005 G8 Presidency (DEFRA, 2005). The Dialogue has engaged G8 and other interested countries with significant energy needs. It has focused on:

■ the strategic challenge of transforming our energy systems to create a more secure and sustainable future
■ monitoring implementation of the commitments made in the associated Gleneagles Plan of Action
■ sharing best practice between participating governments.

The Japan–UK Low-Carbon Society project has contributed to the first and third of these objectives. It took as its starting point the need to stabilize greenhouse gas concentrations at a level that would avoid dangerous climate change. It then went on to create visions of low-carbon societies, identifying the concrete steps required to achieve the necessary transitions.

The two governments have worked with three of the top climate and energy research centres in Japan and the UK – the National Institute for Environmental Studies (NIES), the Tyndall Centre on Climate Change, and the UK Energy Research Centre (UKERC). The Centres undertook a sequence of three workshops and symposia, involving both researchers and stakeholders from a diverse group of some 20 developed and developing countries.

A major component of the project was an international modelling comparison exercise, 'Low-Carbon Society (LCS) Scenarios Towards 2050', undertaken by nine national teams, with a strong developing-country focus. Core model runs were a *Base case*, a *Carbon price* case (rising to $100/tCO$_2$ by 2050), and a '*Carbon-plus*' case to investigate an LCS with a 50% reduction in global CO$_2$ emissions by 2050. This was the level of global emissions reduction referred to in the outcomes of the Heiligendamm G8 Summit in June 2007 (Federal Government of Germany, 2007). The comparison focused on individual model strengths (notably technological change, international emissions trading, non-price mechanisms relating to sustainable development, and behavioural change) rather than a common integrated assumption set. The bulk of this *Climate Policy* supplement is devoted to reporting the outcomes of the international modelling comparison exercise. Strachan et al. (2008) present an overview of the exercise and synthesize the key conclusions. Other papers describe in more detail the conclusions for individual countries.

■ *Corresponding author. E-mail*: jim.skea@ukerc.ac.uk

CLIMATE POLICY 8 (2008) S5–S16

doi:10.3763/cpol.2008.0487 © 2008 Earthscan ISSN 1469-3062 (print), 1752-7457 (online) www.climatepolicy.com

earthscan

The purpose of this editorial is to set the international modelling comparison exercise in the context of the wider Japan–UK Low-Carbon Society project and to describe the conclusions arrived at during the course of the two-year project. This article covers: a working definition of the low-carbon society concept; the need for, and feasibility of, achieving low-carbon societies; establishing and developing low-carbon society visions; evidence of the scope for action offered by existing initiatives at the country, city and sectoral level; the roles of business, the investment community, technology, city authorities and consumers; aligning low-carbon societies with wider sustainable development needs; and policy recommendations.

What is a low-carbon society?

At the first project workshop in June 2006, the Steering Committee was challenged to set out its views on what, in concrete terms, would constitute a low-carbon society. The following working definition was proposed. This was not intended as a scientific statement but rather as a flexible framework which would allow fruitful discussions, leading to practical actions. The definition was intended to capture the perspectives and needs of countries at all stages of development. A consensus was reached at the first workshop that this definition did indeed provide a basis for research and action (National Institute for Environmental Studies, 2006, pp. ii–iii).

A low-carbon society should:

- take actions that are compatible with the principles of sustainable development, ensuring that the development needs of all groups within society are met
- make an equitable contribution towards the global effort to stabilize the atmospheric concentration of CO_2 and other greenhouse gases at a level that will avoid dangerous climate change, through deep cuts in global emissions
- demonstrate a high level of energy efficiency and use low-carbon energy sources and production technologies
- adopt patterns of consumption and behaviour that are consistent with low levels of greenhouse gas emissions.

Although the definition is intended to cover all national circumstances, the implications are different for countries at different stages of development. For developed countries, achieving a low-carbon society would involve making deep cuts in CO_2 emissions by the middle of the 21st century. It would involve the development and deployment of low-carbon technologies and changes to lifestyles and institutions. For developing countries, the achievement of a low-carbon society must go hand in hand with the achievement of wider development goals. This would be with a view to ultimately achieving an advanced state of development, with CO_2 intensity commensurate with that achieved by low-carbon societies in developed countries.

Another key feature is that the definition, while not neglecting the role of technology in any way, also emphasizes the importance of lifestyle and social change. This, along with the close link between the low-carbon society concept and that of sustainable development more broadly, was to become a defining feature of the project as it progressed.

Methods

The project was launched by the then Japanese Environment Minister, Yuriko Koike, and the British Ambassador to Japan, Graham Fry, in Tokyo on 16 February 2006.[1] This was followed by three major international events in Tokyo, London, and then once again Tokyo in June 2006, June 2007 and February 2008, respectively. In each case a one-day symposium open to a wide range of stakeholders was linked to an intense two-day workshop involving low-carbon society researchers from a range of countries.

The goals of the first workshop in June 2006 were to:

a) identify and understand the need for deep cuts in greenhouse gas (GHG) emissions towards 2050 based on scientific findings
b) review country-level GHG emissions scenario studies in developed and developing countries
c) align sustainable development and climate objectives
d) study methodologies to achieve low-carbon societies
e) identify the gaps between our goals for developing country-level low-carbon society scenarios and the current reality
f) identify opportunities for cooperation and how best to cooperate in estimating country, regional and global-level, low-carbon society scenarios (National Institute for Environmental Studies, 2006).

Out of objectives (b) and (d) came the idea for the international modelling comparison exercise which was discussed in greater depth at a technical workshop in Oxford in December 2006.

The second workshop, in June 2007, adopted two broad goals:

1. to demonstrate and raise awareness of the benefits of transitioning to a low-carbon society through sustainable development
2. to develop recommendations on how to close the gap between the business-as-usual and low-carbon society scenarios.

The first goal was addressed by: demonstrating that a low-carbon society could be consistent with policies relating to the environment, economy, development, access to energy and energy security; involving a wider range of stakeholders (including business leaders, policymakers, academics and NGOs) to assist with raising awareness of the low-carbon society concept and disseminating low-carbon society information; to provide expert input on the practicalities of transitioning to a low-carbon society; and sharing expertise and further building analytical capacity relating to low-carbon society visions and modelling.

The second goal was addressed by: identifying feasible contributions that large sectors could make in achieving a low-carbon society; exploring what low-carbon cities might look like and showcasing existing examples; and drafting policy options to achieve a low-carbon society with reference to timeframes and the need for swift action.

A third and final workshop was held in Tokyo in February 2008. This developed key findings and recommendations in four areas: behaviour change and its impact on delivering low-carbon societies; delivering low-carbon societies through sustainable development; enabling low-carbon societies through investment; and addressing opportunities and barriers to a low-carbon society, especially in economically sensitive sectors. The workshop was followed by an open symposium where the conclusions of the workshop series were presented to a wider stakeholder audience including government, business and NGOs.

The feasibility of achieving a low-carbon society

The first workshop included a comprehensive set of presentations and papers describing emissions scenario exercises both globally and for a range of developed and developing countries (National

Institute for Environmental Studies, 2006). The countries included Japan, the UK, France, Germany, the European Union, Canada, Russia, China, Mexico, India, Brazil, South Africa and Thailand. A key focus for the discussions was the various methodological approaches used by the modelling teams. These discussions led directly to the establishment of the international modelling comparisons project described in this *Climate Policy* supplement.

The key message from both the workshop presentations and the subsequent modelling activity was that the achievement of low-carbon societies, in the context of global CO_2 emission reduction of 50% by 2050, is indeed feasible in technological and economic terms. Energy efficiency, demand-side responses and the choice of technologies for electricity generation were among the most important contributors to emissions reduction. Novel transport technologies such as fuel-cell hydrogen vehicles and plug-in hybrids also played an important role in some model runs.

For such a 50% global CO_2 emission reduction, most models in this LCS project comparison showed an associated GDP loss in the range 0.35–1.35% annually by 2050, though one model showed an increase in GDP due to the stimulus provided by higher levels of investment in low-carbon technologies.[2] However, the required carbon price signal or marginal cost of abatement was found to be in the range $100–330/tCO_2$. This greatly exceeds the current price of carbon in the EU Emissions Trading Scheme, which was trading at just over $30/tCO_2$ in early March 2008 (Point Carbon, 2008). There is a serious question about the political viability of establishing a market signal, through taxation or otherwise, at these higher levels right across the economy. There is also a set of issues regarding international cooperation in emissions reductions, including flexible and equitable linking of emissions markets. So although modelling activity has demonstrated that low-carbon societies are technically and economically feasible, there is a major challenge in putting in place policies that will secure the technological and behavioural changes required.

Visions of a low-carbon society

A key element of turning the possibility of a low-carbon society into a reality is to develop visions that will be credible and attractive to the general public. A good example of this is the Japan Low-Carbon Society project. This envisages a world in which global temperature rise is held below 2°C, global CO_2 emissions are cut by 50% by 2050, and Japanese emissions are cut by 70%. The results of this project were presented at Workshop 2 in London (Matsuoka, 2007; National Institute for Environmental Studies et al., 2007).

However, a key conclusion is that more than one path of social development is consistent with this outcome. The research team constructed two contrasting visions of a Japanese low-carbon society.[3] Vision A (*'Doraemon'*)[4] is technology-driven, with citizens placing great emphasis on comfort and convenience. They live urban lifestyles with centralized production systems and GDP per capita growing at about 2% per annum. Vision B (*'Satsuki and Mei'*)[5] is of a slower-paced, nature-oriented society. People tend to live in decentralized communities that are self-sufficient in that both production and consumption are locally based. This society emphasizes social and cultural values rather than individual ambition.

In both cases a 70% reduction in CO_2 emissions is achieved by 2050. However, the mix of technologies employed is different (Table 1). In both cases, energy efficiency improves considerably – both in industry and in the home. The big differences are in: transport needs relating to the very different patterns of settlement; and in the structure of electricity production. The technology-driven society relies heavily on nuclear power and fossil-fuel use coupled with carbon capture and storage. Hydrogen is produced for use in fuel-cell vehicles. The nature-oriented society instead relies heavily on biomass, both for electricity generation and for biofuel production for use in hybrid vehicles.

TABLE 1 Comparison of CO_2 emission reduction drivers

	Vision A ('*Doraemon*')	Vision B ('*Satsuki and Mei*')
Society	High economic growth Decrease in population and number of households	Reduction of final demand by material saturation Reduction in raw material production Decrease in population and number of households
Industrial	Energy-efficient improvement of furnaces and motors etc. Fuel-switching from coal/oil to natural gas	Energy-efficient improvement of furnaces and motors etc. Increase in fuel-switching from coal and oil to natural gas and biomass
Residential and commercial	High-insulation dwellings and buildings Home/building energy management system Efficient air-conditioners Efficient water heaters Efficient lighting systems Fuel-cell systems Photovoltaics on the roof	High-insulation dwellings and buildings Eco-life navigation system Efficient air-conditioners Efficient water heaters Efficient lighting systems Photovoltaics on the roof Expanding biomass energy use in home Diffusion of solar water heating
Transportation	Intensive land use Concentrated urban function Public transportation system Electric battery vehicles Fuel cell battery vehicles	Shortening trip distances for commuting through intensive land use Infrastructure for pedestrians and bicycle riders (sidewalk, bikeway, cycle parking) Biomass–hybrid engine vehicle
Energy Transformation	Nuclear energy Effective use of electricity in night time with storage Hydrogen supply with low- carbon energy sources Advanced fossil-fuelled plants + carbon capture and storage Hydrogen supply using fossil fuel + carbon capture and storage	 Expanding share of both advanced gas combined cycle and biomass generation

Source: Matsuoka (2007).

Both paths of social development will lead towards a low-carbon society. More realistically, a variety of development paths that combine elements of these more extreme visions would also be compatible with deep reductions in CO_2 emissions, demonstrating the robustness of the low-carbon society goal.

Practical actions for a low-carbon society

While low-carbon societies are a long-term goal, there are practical steps that can be taken today to put ourselves on the right trajectory. The second low-carbon society workshop focused on a number of case studies where concrete steps were already under way, mainly at the city level, to realize the low-carbon vision.

Watson (2007) described how Arup, the global design and consulting firm, is collaborating with the Shanghai Industrial Investment Corporation to plan the new city of Dongtan in China, covering an area three-quarters the size of Manhattan at the mouth of the Yangtze River. Dongtan will have low energy consumption and will be as close to carbon-neutral as is possible. The development will include electricity generation from wind, solar, biofuel and recycled city waste, with hydrogen fuel cells being used to power public transport. A network of cycle and footpaths will help the city achieve close to zero vehicle emissions. A key part of the initiative will be ongoing research and evaluation activity which will ensure that lessons are learned and can be applied in subsequent developments.

Shimada (2007) focused on the sustainability challenge in Shiga prefecture, Japan. Here, the challenge is to restore water quality in Lake Biwa, reduce the volume of waste going to landfill by 75% and reduce CO_2 emissions by 50% by 2030. The plan involves a three-way partnership between citizens, business and local government. The goal is that the partners will share economic benefits as well as environmental gains through 'sustainable taxation' and 'sustainable finance'. Specific measures would include environmental regulations, regulations on land use and construction, subsidies for advanced technologies, voluntary environmental action plans, and awareness/education programmes. This case study highlighted the central role that governments must play in taking forward the low-carbon society concept.

Deacon (2007) outlined the comprehensive actions being taken by the Mayor of London to cut CO_2 emissions across the city. The Action Plan sets targets for 2025 and comprises four main programmes: Green Homes, Green Organizations, Green Energy and Green Transport. The Green Homes programme could cut CO_2 emissions in this sector by almost half, helped by subsidies for house insulation and energy-efficient devices. The Green Organizations programme aims to encourage businesses to cut their energy use through simple managerial approaches – turning off lights and IT equipment – and improving building energy efficiency. The ambitious goal of the Green Energy programme is to take one-quarter of London's electricity supply off the National Grid and meet it through more efficient local energy systems. The Green Transport programme will encourage people on to public transport, using methods such as congestion charging, and will incentivize more fuel-efficient vehicles, for example by exempting them from congestion and parking charges. Altogether, London's CO_2 emissions should fall by 60% by 2025 compared with 1990 levels.

The role of stakeholder groups

The second workshop also squarely addressed the role that different stakeholders' groups could play in developing low-carbon societies. The workshop addressed the roles of the investment community, business and consumers. However, a key message from the associated stakeholder symposium was that government had the primary responsibility to initiate low-carbon development (DEFRA, 2007, p.34). Businesses, for example, can deliver results, but governments must provide the frameworks and incentives to direct business activities in the right direction.

The world will be *investing* around US$20 trillion between 2005 and 2030 in energy infrastructure, with developing countries making up on slightly more than half of this spending. China alone will make up almost one-third of the total from developing countries (Garibaldi, 2007). An alternative scenario modelled by the IEA (IEA, 2006) shows that emissions reductions of 16% or 6.3 billion tCO_2 by 2030 can be achieved without incurring additional investment relative to the IEA reference scenario.

Delivering these potential emission reductions requires shifting part of the total energy sector investment from supply expansion to energy efficiency and the demand side. In addition, public and private energy technology R&D investment has been decreasing over the past decades, due partly to liberalization in the energy sector, the lack of clear long-term signals, and an inappropriate regulatory environment. This trend needs to be reversed.

Current mechanisms that leverage financing play an important, but so far insufficient, role in promoting low-carbon investment. Clean Development Mechanism (CDM) projects in the pipeline to 2012 are expected to generate 2 billion tCO_2e of certified emissions reductions worth perhaps $4 billion p.a. (Acquatella, 2007). However, this is only a fraction of the estimated $20–30 billion p.a. of low-carbon investment required in developing countries. The removal of barriers, such as high transaction costs, technology risk and policy uncertainty, could enable an increase in the scale of investment through CDM. The Global Environment Facility (GEF) has provided US$6.2 billion in grants and has generated over US$20 billion in co-financing from other sources since 1991. According to the World Bank, a tenfold increase would be needed to finance a strategic global programme to support cost reduction of pre-commercial low-carbon technologies. Trading on the EU Emissions Trading Scheme (EU ETS) is now worth US$8 billion per year. The global carbon market shows a very large growth potential, especially if the largest global emitters join in carbon trading and if emerging regional carbon trading schemes merge into a single market.

Expanding capacity for demand-side investment will require innovation and structural change within the finance sector. Most energy finance facilities are ill prepared to handle barriers associated with energy efficiency programmes. These include financial constraints on individual consumers, high implicit discount rates, partial information on energy performance of end-use appliances, the need to organize a large number of individual actions, and partial information on the potential savings to demand-side investment.

To support the further expansion of financing mechanisms for low-carbon investment, clear, stable, long-term signals are needed, particularly in terms of a global price for carbon. Market mechanisms, such as carbon taxation and emissions trading, are key. Energy subsidies and tariff barriers need to be dismantled.

Business is now increasingly of the view that there is no inherent conflict between having a healthy, competitive economy and a cleaner environment. Access to mobility, illustrated for example by the '*Doraemon*' vision cited above, can be reconciled with a low-carbon society through the continuous application of innovation and creativity (Smith, 2007).

The transport arena serves as a good example of a sector in which business has a big role to play. But there are no 'silver bullets' to reduce greenhouse gas emissions. In transport, there are three broad levels at which steps can be taken to reduce greenhouse gas emissions. The first is simply through human behaviour. Sticking to speed limits would help, as would the adoption of smarter driving techniques which can improve vehicle efficiency by close to 10%. With nearly one-third of car journeys involving a distance of less than 2 miles, there is ample opportunity to use alternative modes of transport such as walking, cycling and public transport. Good infrastructure and planning policies designed to reduce the need to travel can also have a longer-term effect. The UK Low-Carbon Vehicle Partnership (LowCVP) has argued strongly for information and education campaigns to encourage low-carbon vehicle purchase and smarter driving.

The second level is the development of cleaner vehicle technologies and fuels. Hybrid petrol–electric vehicles are already beginning to have an impact on the market. These can enable a pathway to more radically advanced vehicle designs based on alternative fuels (such as hydrogen), alternative means of propulsion (fuel cells), and electric vehicles. Vehicle fuels with a lower carbon footprint, including sustainably produced hydrogen and biofuels will also play their part.

The final level is through changes to the transport system itself. As well as modal switching, it is possible to imagine that information and communication technologies can be used to inform drivers so that congestion is avoided and CO_2 emissions are consequently reduced. Taking these three levels together – human behaviour, vehicle technology and traffic systems – it is possible to envisage the provision of a given level of mobility service with 70% less CO_2 emissions than at present (Yamashita, 2008).

There has been a growing acknowledgment that the way *consumers* approach energy must change if rapid decreases in greenhouse gas emissions are to be achieved. Economics and psychology have dominated consumption research. However, a purely economic perspective is not enough, as individuals do not operate in a social vacuum. Equally, psychology focuses on the individual and downplays the importance of the social and cultural contexts in which energy is used. In fact, energy consumption is structured by the range of choices available (Jackson, 2007; Wilhite, 2007). Energy demand can be conceptualized as a product of both choosers and the set of choices available. No amount of taxes or price increases on automobiles or fuel will affect mobility practices unless the consumer has an alternative. Air conditioning is growing rapidly around the world, not because people have elevated their demands for thermal comfort, but because the constructed world that they inhabit no longer allows for comfortable natural cooling. Energy policy needs to broaden its focus from technical and market efficiency to an examination of how energy service needs can be achieved in the least energy-intensive way. Policies aiming at deep reductions in CO_2 must take a holistic and long-term approach to change.

Low-carbon societies and sustainable development

It is often assumed that there must be trade-offs between environmental quality and socio-economic development. However, an important conclusion of the Low-Carbon Society project was that pathways to achieve developmental goals can be climate-friendly and that sustainable development can be a driving force for addressing climate challenges. Rather than thinking in terms of trade-offs, the development/climate 'frontier' can be pushed back through technological and institutional innovation, international and regional cooperation, targeted technology and investment flows, and working to align stakeholder interests (National Institute for Environmental Studies, 2006, p. 39).

Given both the growth and greenhouse gas mitigation potential in developing countries, it is more efficient to focus on the co-benefits between sustainable development and the pursuit of a low-carbon society. One approach is for developing countries to make pledges to implement sustainable development policies and measures (SD-PAMS). Starting from development objectives, countries would map out the implementation of policies and measures in a manner that would take cognizance of the need to mitigate or adapt to climate change.

Each individual country will take its development path on the basis of its local resource endowments. It is therefore important to focus on the potential barriers and incentives that can help developing countries in moving towards a lower carbon future. The workshops showcased efforts by various developing countries which have created good examples of win–win strategies associated with emission reduction efforts.

An *Indian* case study (Shukla, 2006) showed that energy policies framed on India's sustainable development vision could well align energy security and a low-carbon future.[6] Under most scenarios, per capita CO_2 emissions would remain low, compared to global averages, throughout the 21st century while, at the same time, India can further contribute to cost-effective greenhouse gas stabilization, offering low-cost mitigation opportunities, given the appropriate incentives. There would be substantial co-benefits associated with the joint reduction of SO_2 and CO_2 emissions, although realizing this would require modifications to international instruments such as the CDM and greater alignment between national and global environmental regimes in general. Again, in India, electricity reforms have reduced the carbon content of electricity compared with the baseline.

In *Brazil*, measures relating to land-use change and energy have a large potential to reconcile greenhouse gas mitigation and sustainable development (La Rovere, 2006). The major challenge is to limit deforestation, which has drivers that go far beyond the purely economic domain. Improved governance may increase the enforcement of existing laws and regulations to avoid illegal deforestation in the Amazon region, and thus reduce emissions. However, in the medium and long term, CO_2 emissions from fossil fuel combustion will be the most important factor. The main opportunities to align climate change and sustainable development objectives in Brazil include: energy efficiency in industry and transport; the greater use of natural gas in the industrial, residential and commercial sectors; greater exploitation of hydropower potential; the production of ethanol from sugar cane for use as a vehicle fuel; blending biodiesel with diesel oil to fuel buses and trucks; and renewable power generation to promote wider access to electricity among the rural population.

South Africa's development objectives focus on growth, job creation, and access to key services including energy and housing (Mwakasonda, 2006). Increasing the percentage of renewable energy in the electricity generation mix is a specific goal. The government strategy aims to generate 5% of the national grid-supplied power from renewable technologies, such as micro-hydro, biomass-fuelled turbines, solar thermal, wind turbines and PV. A national target for renewable energy sources can lead to local environmental benefits, and GHG reductions. At least 50% of all new houses built in communities incorporate climate-conscious solar passive design principles in their construction, thereby eliminating the need for space heating and cooling, resulting in lower CO_2 emissions.

Efforts at the national level must be supported by international action. Making low-carbon technologies and finance available to developing countries is a key measure. It is also necessary to spell out the incentives for sustainable development and climate policies in formulating support programmes.

Conclusions and recommendations

The Low-Carbon Society project has shown that low-carbon societies are achievable, but also that it will require a major coordinated effort, nationally and internationally, to achieve this vision. Although advancing the technological frontier will be vital, changes must go to a deeper social level if climate change and development goals are to be reconciled.

The key conclusions emerging from the workshop series were that:

▪ Achieving the transition to a low-carbon society is essential if greenhouse gas concentrations in the atmosphere are to be stabilized at a safe level. Modelling and scenario work has shown that this transition is possible.

▦ It will be less costly to move towards a low-carbon society than it will be to delay climate change mitigation efforts and experience the more extreme impacts of climate change.

▦ Long-term certainty is needed to create the market conditions for investment in low-carbon solutions – a comprehensive approach to RD&D for low-carbon technology as well as emerging markets, products and services is required to underpin this investment.

▦ Some of the more substantial changes will be required in the built environment, transport, and power sectors.

▦ There are major synergies between policies that promote sustainable development objectives and those that encourage the transition to a low-carbon society. Pursuing these policies can deliver significant economic, social and environmental co-benefits, especially in developing countries.

▦ The role of government is critical and top-level political leadership will be essential. Governments must establish the enabling conditions under which individuals, business and organizations can benefit from the opportunities offered by new low-carbon markets, technologies, products and services. A portfolio of policies will be required to achieve this.

▦ The building of trust within and between nations is essential to reinforce the credibility of long-term goals and policies.

▦ Consumer choice and individual action, in the context of clear policies that enable low-carbon options and lifestyles, can be powerful drivers in delivering the level of behaviour change required to enable the transition to low-carbon societies.

There are clear inferences to be drawn from these conclusions in terms of policy. If low-carbon societies are to be achieved, action is needed in the following areas:

▦ Long-term goals to reduce global greenhouse gas emissions by at least 50% of 1990 levels by 2050 are required.

▦ Long-term policy signals to strengthen carbon pricing, e.g. through taxation and enhanced international emissions trading, should be established to create appropriate incentives for business.

▦ It would help if tax burdens were shifted away from income and employment towards environmental pollution in order to internalize the cost of CO_2 emissions and encourage businesses and individuals to reduce emissions.

▦ The focus of development investment in developing countries should be shifted towards lower-carbon approaches.

▦ There needs to be a step change in the transfer of low-carbon technologies to developing countries. This can be achieved by expanding financial flows and developing new financing mechanisms.

▦ Trade regimes should be adjusted to encourage rapid deployment of technologies and products that enhance sustainable development while lowering CO_2 emissions.

▦ Energy efficiency improvement should be accelerated, using incentives that encourage institutional and behavioural change.

▦ The demonstration and deployment of near commercial technologies, such as carbon capture and storage, is required, as is significantly increased investment in R&D for technologies with greater promise in the long term.

▦ Policies and frameworks should be implemented which enable a change in human behaviour and lifestyle, by removing high-carbon choices and providing consumers with the opportunity to benefit from low-carbon approaches.

▧ The required level of trust can only be built by continuing and enhancing dialogue between stakeholder groups within countries and between countries with diverse national circumstances. International cooperation should be enhanced, as should the sharing of expertise and best practice between national, regional and international stakeholders.

None of this will be easy. Much further work will be required to articulate policy measures in the necessary detail. But the Low-Carbon Society project has demonstrated that this task is both necessary and possible. It will also bring benefits in terms of human development that go beyond the climate arena.

Acknowledgements

The authors acted as co-chairs of the meetings associated with the Low-Carbon Society project. Any views expressed are those authors alone and do not necessarily reflect the views of other members of the project Steering Committee, other workshop and symposium participants, or the governments of Japan and the UK.

Notes

1. This was the first anniversary of the Kyoto Protocol having come into effect.
2. See Strachan et al. (2008) for a discussion of the broad uncertainties in drivers of this cost range.
3. These visions are discussed and quantified in Fujino et al. (2008).
4. *Doraemon* is a Japanese comic series about a robotic cat who travels back in time from the 22nd century.
5. Satsuki and Mei are the daughters in the film '*My Neighbor Totoro*'. They live an old house in rural Japan, near which many curious and magical creatures live.
6. This is also discussed and quantified in Shukla et al. (2008).

References

Acquatella, J., 2007, 'How finance can enable a low-carbon society', in: DEFRA (ed), *Achieving a Low-Carbon Society: Symposium and Workshop, 13–15 June 2007*, DEFRA, London.

Deacon, A., 2007, 'Low-carbon cities', in: DEFRA (ed), *Achieving a Low-Carbon Society: Symposium and Workshop, 13–15 June 2007*, DEFRA, London.

DEFRA, 2005, *Gleneagles Plan of Action: Climate Change, Clean Energy and Sustainable Development*, DEFRA, London [available at www.defra.gov.uk/environment/climatechange/internat/pdf/gleneagles-planofaction.pdf].

DEFRA, 2007, *Achieving a Sustainable Low-Carbon Society: Symposium and Workshop, 13–15 June 2007*, DEFRA, London [available at www.ukerc.ac.uk/Downloads/PDF/07/0706LCS/0706LCSFinalReport.pdf].

Federal Government of Germany, 2007, *Chair's Summary*, G8 Summit, Heiligendamm, 8 June 2007 [available at www.g-8.de/Content/EN/Artikel/__g8-summit/anlagen/chairs-summary,templateId=raw,property=publicationFile.pdf/chairs-summary].

Fujino, J., Hibino, G., Ehara, T., Matsuoka, Y., Masui, T., Kainuma, M., 2008, 'Back-casting analysis for 70% emission reduction in Japan by 2050', *Climate Policy* 8, Supplement, 2008, S108–S124.

Garibaldi, J.A., 2007, 'Catalysing a shift in the development pattern: a programmatic approach', in: DEFRA (ed), *Achieving a Low-Carbon Society: Symposium and Workshop, 13–15 June 2007*, DEFRA, London.

International Energy Agency (IEA), 2006, *Energy Technology Perspectives: Scenarios and Strategies to 2050*, Paris, France.

Jackson, T., 2007, 'Behavioural change', in: DEFRA (ed), *Achieving a Low-Carbon Society: Symposium and Workshop, 13–15 June 2007*, DEFRA, London.

La Rovere, E.L., 2006, 'Aligning climate change and sustainable development objectives in Brazil', in: National Institute for Environmental Studies (ed), *Developing Visions for a Low-Carbon Society (LCS) through Sustainable Development*, Tsukuba. Japan.

Matsuoka, Y., 2007, 'How to link modelling and practical steps to achieve a low-carbon society', in: DEFRA (ed), *Achieving a Low-Carbon Society: Symposium and Workshop, 13–15 June 2007*, DEFRA, London.

Mwakasonda, S., 2006, 'Aligning climate change and sustainable development objectives in South Africa', in: National Institute for Environmental Studies (ed), *Developing Visions for a Low-Carbon Society (LCS) through Sustainable Development*, Tsukuba, Japan.

National Institute for Environmental Studies, 2006, *Developing Visions for a Low-Carbon Society (LCS) through Sustainable Development*, Tsukuba, Japan [available at http://2050.nies.go.jp/material/2050WS-WorkshopSummary_Final.pdf].

National Institute for Environmental Studies (NIES), Kyoto University, Ritsumeikan University, Mizuho Information and Research Institute, 2007, *Japan Scenarios towards Low-Carbon Society (LCS): Feasibility Study for 70% CO_2 Emission Reduction by 2050 below 1990 Level*, NIES, Tsukuba, Japan [available at http://2050.nies.go.jp/interimreport/20070215_report_e.pdf].

Point Carbon, 2008, *EUA OTC Closing Price* [available at www.pointcarbon.com/].

Shimada, K., 2007, 'Shiga prefecture', in: DEFRA (ed), *Achieving a Low-Carbon Society: Symposium and Workshop, 13–15 June 2007*, DEFRA, London.

Shukla, P.R., 2006, 'Perspective, framework and illustration from India', in: National Institute for Environmental Studies (ed), *Developing Visions for a Low-Carbon Society (LCS) through Sustainable Development*, Tsukuba, Japan.

Shukla, P.R., Dhar, S., Mahapatra, D., 2008, 'Low-carbon society scenarios for India', *Climate Policy* 8, Supplement, 2008, S156–S176.

Smith, G., 2007, 'The contribution of business to a low-carbon society', in: DEFRA (ed), *Achieving a Low-Carbon Society: Symposium and Workshop, 13–15 June 2007*, DEFRA, London.

Strachan, N., Foxon, T., Fujino, J., 2008, 'Policy implications from the Low-Carbon Society (LCS) modelling project', *Climate Policy* 8, Supplement, 2008, S17–S29.

Watson, J., 2007, 'Dongtan Eco-city', in: DEFRA (ed), *Achieving a low-carbon society: Symposium and Workshop, 13–15 June 2007*, DEFRA, London.

Wilhite, H., 2007, 'Changing energy consumption: why it happens and how it can be made to happen', in: DEFRA (ed), *Achieving a Low-Carbon Society: Symposium and Workshop, 13–15 June 2007*, DEFRA, London.

Yamashita, M., 2008, 'Roadmap to a low-carbon world: business perspectives', paper presented at the symposium *Roadmap to the Low-Carbon World*, Hotel Metropolitan Edmont, Tokyo, 15 February 2008.

Policy implications from the Low-Carbon Society (LCS) modelling project

NEIL STRACHAN[1]*, TIM FOXON[2], JUNICHI FUJINO[3]

[1] Department of Geography, King's College London, UK
[2] Sustainability Research Institute, School of Earth and Environment, University of Leeds, UK
[3] National Institute for Environmental Studies, Japan

Under the Japan–UK research project 'Low-Carbon Society (LCS) Scenarios Towards 2050', an international modelling comparison was undertaken by nine national teams, with a strong developing-country focus. Core model runs were a *Base case*, a *Carbon price* case (rising to $100/tCO$_2$ by 2050) and a *Carbon-plus* case to investigate an LCS scenario with a 50% reduction in global CO$_2$ emissions by 2050. The comparison emphasis was to focus on individual model strengths (notably technological change, international emissions trading, non-price (sustainable development) mechanisms and behavioural change) rather than a common integrated assumption set. A complex picture of long-term LCS scenarios comes from the range of model types and geographical scale (global vs. national); however, common themes for policy makers do emerge. A core finding is that LCS scenarios are technologically feasible. However, preferred pathways require clear and early target setting and incorporation of emissions targets across all economic activities. This will probably entail significant socio-economic changes. To realize major LCS transitions requires sustained progress in R&D and deployment of a broad range of technologies, with carbon capture and storage (CCS) a key technology in most low-carbon portfolios. Developing countries, in particular, face an immense challenge to achieve LCS in light of their economic growth requirements. As such, international cooperation is required in iterative and flexible burden sharing under international emissions trading regimes.

Keywords: carbon pricing; climate change; climate stabilization; energy models; low-carbon society; mitigation; public policy; scenario modelling; stabilization pathways

Dans le cadre du projet de recherche Japon-Royaume-Uni de scénarios de société sobre en carbone à l'horizon 2050 "Low-Carbon Society (LCS) Scenarios Towards 2050", une comparaison internationale des modélisations fut entreprise par neuf équipes nationales, avec une forte représentation des pays en développement. Les simulations principales comprenaient un « cas de référence », un cas « prix du carbone » (allant à $100/tCO$_2$ d'ici 2050) et un cas « carbone plus » pour l'examen d'une LCS de 50% de réduction des émissions mondiales de CO$_2$ d'ici 2050. L'accent de la comparaison fut porté sur les forces individuelles de chaque modèle (notamment le changement technologique, l'échange international d'émissions, les mécanismes hors-prix (développement durable), et les changements de comportement), plutôt que sur une série intégrée d'hypothèses collectives. Une image complexe de scénarios LCS à long-terme émerge des différents types de modèles et de l'échelle géographique (mondiale ou nationale), cependant des thèmes communs pour les décideurs font sont révélés. Une conclusion clé est la faisabilité technique des scénarios LCS. Cependant il est nécessaire de mettre en place les objectifs de manière claire et assez tôt pour favoriser les axes préférés ainsi que l'intégration d'objectifs d'émissions à travers toutes les activités économiques. Ceci entraînera probablement des changements socio-économiques. Dans le but d'obtenir de larges transitions LCS, un progrès soutenu est nécessaire en recherche et développement et par le déploiement d'un large éventail de technologies, la capture et le stockage de carbone (CSC) étant une technologie clé dans la plupart des portefeuilles sobres en carbone. Les pays en développement en particulier font face à un défi énorme quant à la réalisation de LCS du fait de

■ *Corresponding author. E-mail*: neil.strachan@kcl.ac.uk

CLIMATE POLICY 8 (2008) S17–S29

doi:10.3763/cpol.2007.0488 © 2008 Earthscan ISSN: 1469-3062 (print), 1752-7457 (online) www.climatepolicy.com

leurs besoins en croissance économique. Ainsi, une coopération internationale est nécessaire selon un « burden sharing » itératif et flexible dans le cadre d'un régime international d'échange de quotas.

Mots clés: changement climatique; fixation du prix du carbone; mitigation; modèles énergétiques; modélisation de scenarios; politique publique; société sobre en carbone; stabilisation du climat; voies de stabilisation

1. Introduction to the Low-Carbon Society (LCS) research project

This article draws out the main policy implications from the findings of the low-carbon society (LCS) scenarios investigated by nine international modelling teams under the Japan–UK research project 'Low-Carbon Society Scenarios Towards 2050'. Each modelling team was asked to run a *Base case*, a *Carbon price* case (rising to $100/tCO$_2$ by 2050), and one or more *Carbon-plus* cases to investigate a LCS scenario with a 50% reduction in global CO$_2$ emissions by 2050.

The latest scientific consensus (IPCC, 2007a) has further strengthened the evidence that it is very likely that increased warming over the last half-century is due to anthropogenic emissions of greenhouse gases (GHGs) and that continued GHG emissions at or above current rates would cause further warming and induce many changes in the global climate system during the 21st century. A major recent report on the economics of global climate change (Stern, 2006) supported the position that the benefits of stringent climate mitigation action outweigh the costs and risks of delayed action. Notwithstanding the high profile of the *Stern Review*, extensive debate continues over the costs of damages from global climate change (Pearce, 2003) and the complexities in valuing both costs and benefits of climate change policies, notably in the roles of long time scales and pervasive uncertainty (Nordhaus, 1994). Key issues include the drivers of baseline emissions, the development of new technologies, and the role of behavioural change in emissions reductions (IPCC, 2007b). In addition, a large literature exists on the moral responsibility and political feasibility of designing a coordinated global response to climate change (Barrett, 2003).

The international policy response has focused on the United Nations Framework Convention on Climate Change (UNFCCC), with the Kyoto Protocol entering into force in 2005 as a first, albeit limited, step on mitigation pathways. The Kyoto Protocol had modest reduction targets only up to 2012, and omitted major emitter countries. However, major policy initiatives, notably the EU Emissions Trading Scheme (EU ETS), were set up in response to Kyoto targets. At the UNFCCC Conference in Bali, December 2007, a roadmap and agenda were agreed for achieving, by the end of 2009, a new deal for commitments beyond 2012.

A complementary mechanism is the G8 Gleneagles Dialogue, begun under the UK's G8 presidency in 2005. This process brings together 20 countries with the greatest energy demands, including the G8 and the major emerging economies of Brazil, China, India, Mexico and South Africa, to discuss innovative ideas and new measures to tackle climate change. In addition, the G8 have engaged with the World Bank and other international financial institutions to create a new investment framework for clean energy and development, including investment and financing, as well as commissioning alternative energy scenarios and strategic analysis (IEA, 2006). The Japanese government agreed to receive a report on the Gleneagles Dialogue during their G8 presidency in 2008 and launched a new initiative called 'Cool Earth 50' on 24 May 2007, which aims for 50% reductions in global GHG emissions by 2050 through developing innovative technologies and building a low-carbon society.

The modelling effort described here takes its lead from the declaration issued during the 2007 G8 summit in Heiligendamm (G8 Communiqué, 2007) supporting a global target of a 50%

reduction in GHGs by 2050. This corresponds with the more stringent stabilization pathways envisioned under the IPCC report (IPCC, 2007b), which, so far, have been subject to relatively little scenario analysis (only 24 out of 177 scenarios). Individual countries have led on long-term target setting, with the UK currently codifying its legally binding long-term national CO_2 reduction target of 60% by 2050 through the Climate Change Bill (DEFRA, 2007).

There is thus a considerable need to investigate and understand what long-term global mitigation targets mean for individual countries and sectors, and the importance and interactions between technological and behavioural measures to reduce emissions. A further important interaction is with emission reductions and broader considerations of sustainable economic development. To this end, in 2006, Japan and the UK launched the joint research project 'Low-Carbon Society Scenarios Towards 2050' to deepen our understanding of the need to reduce greenhouse gases in order to achieve low-carbon societies (LCS).

The Japan–UK Low-Carbon Society project is discussed in depth in another paper in this supplement (Skea and Nishioka, 2008), which sets the policy context for this international modelling comparison exercise. Three workshops were held as part of this project – Tokyo, February 2006; London, June 2007; and Tokyo, February 2008. Through this process, a working definition for the LCS concept was established; the need for achieving LCS was established; feasible and equitable LCS scenarios were developed; evidence of the scope for practical action at the country, city and sectoral level were collated; the critical role of stakeholders – business, the investment community, technology vendors, local government and consumers – was highlighted; and, finally, the necessary alignment of low-carbon societies with wider sustainable development needs was emphasized. Skea and Nishioka (2008) present final conclusions and policy recommendations.

A key component of the Japan–UK LCS research project and focused outcome was a collaborative international modelling comparison. This article describes the results of this innovative process undertaken by nine national teams, with strong developing-country participation, emphasizing long-term deep reductions in CO_2 and other GHGs, and with complementary model runs based on common LCS scenarios.

2. Overview of the LCS modelling studies

The comparative modelling initiative under the Japan–UK LCS research project was initiated at the UKERC Annual Energy Modelling Conference (AEMC), held in Oxford, UK, in December 2006.[1] This was an open symposium with UK energy policy stakeholders followed by a technical modelling workshop. The principal outcome of the AEMC was an agreement to use models of different scales and types to investigate the restructuring of energy systems under long-term LCS scenarios. This involves the investigation of technology pathways, behavioural responses, economic implications and required policy measures.

Unlike earlier collaborative modelling projects (IMCP: Edenhofer et al., 2006; EMF 23: van Vuuren et al., 2006), this LCS project did not run models under common assumptions to investigate a single key issue in energy futures (e.g. induced technological change or multi-gas mitigation). Instead, the LCS modelling project had a broader goal – to identify policy implications from scenarios of long-term deep reductions in CO_2 (and other GHGs), combining findings on economic, technological and behavioural implications.[2] Furthermore, there was an emphasis on developing-country participation,[3] on a mixture of country vs. global models, and on top-down vs. bottom-up models. The advantage of this approach is that individual models can investigate aspects of long-run LCS scenarios based on the particular strength of that approach (listed in the final column of Table 1) and hence highlight a number of policy-relevant insights. The disadvantage is that a rigorous comparison of model outputs is not applicable, with only summary information given in Table 1.

LCS scenarios are described as societies consistent with reducing roughly global emissions of GHGs or CO_2 by 50% by 2050 compared with the base year (2000), whilst satisfying adequate energy service demands, through proposing possible combinations of technological and social innovations based on favourable socio-economic future visions. In order to move from scenarios to modelling, there is a need to quantify future socio-economic scenarios, identify potential measures towards achieving LCS and verify favourable combinations to realize them.

LCS modelling tools quantify and qualify future trends, potentials and actions. The models all have their own strengths and weaknesses, different coverage, different scales and special features. Development and climate need to be aligned for the transition to a low(er)-carbon society, as global development along a high-carbon path is untenable and stand-alone decarbonization is costly.

TABLE 1 Key characteristics and results of LCS models

No.	Modelling paper	Model	Global/ National	Top-down/ bottom-up	*Carbon-plus* $/tCO_2$ and % change in GDP (2050)	Key characteristic of model and/or analysis of *Carbon-plus* run
1	Barker et al. (UK)	E3MG	G	TD	$100/tCO_2$ (by 2030) +1.10%	Technological change, emissions revenues
2	Akimoto et al. (Japan)	DNE21	G	BU	Averaged cost of $45.2–49.6/tCO_2$	International sectoral approach
3	Remme et al. (Germany)	TIMES	G	BU	$330/tCO_2$ −1.3%	Technological change
4	Edmonds et al. (USA)	MiniCam	G	BU	$136/tCO_2$	Technological change, integrated assessment
5	Bataille et al. (Canada)	CIMS	N	Hybrid simulation	$175–200/tCO_2$	Price vs. non-price measures, emissions trading
6	Fujino et al. (Japan)	Linked models	N	Hybrid	−0.83% to −0.90% −0.96% to −1.06%	Feasibility of long-term stringent CO_2 reductions
7	Strachan et al. (UK)	MARKAL-Macro	N	Hybrid	$402–490/tCO_2$ −1.64% to −2.21%	International drivers on UK
8	Shrestha et al. (Thailand)	AIM	N	BU	(runs only up to $100/tCO_2$)	Technological change
9	Shukla et al. (India)	AIM and MARKAL	N	Soft-linked TD/BU	0% to −1.35%	Sustainable development

The set of agreed scenarios for all modelling teams were:

1. *Base case*
   - Existing model base cases were used, linked to IPCC scenarios or country-level forecasts.
2. *Carbon price*: $10/tCO$_2$ in 2013 rising exponentially to $100/tCO$_2$ in 2050.
   - A forecasting run to ascertain the impact of a relatively modest carbon price signal.
3. *Carbon-plus*: as *Carbon price* plus additional measures to achieve long-term LCS.
   - For this project the LCS target is a 50% global reduction in GHG emissions by 2050 to match the 2007 G8 summit statement.
   - This global target provides an 'envelope' for individual model targets.[4]
   - To allow for flexibility between different modelling approaches, the LCS target may be interpreted in terms of equivalent targets for atmospheric stabilization, radiative forcing, or maximum temperature change.

Additional scenarios to explore key uncertainties were encouraged but were optional.

Each modelling team used its existing input parameters and model structure. First, this minimized model set-up, particularly in the base case, and without the necessity for significant model development. Second, by using the individual strengths of each model (notably in the *Carbon-plus* run), alternative policy insights from models were generated, especially between developed and developing countries, and between global and national models.

3. Comparison of results

Table 1 lists the papers and models and presents summary information.[5] The table and resultant discussion is organized first by geographical coverage (global to national), and then by structural type (top-down to hybrid to bottom-up). The final two columns focus on the *Carbon-plus* runs and list CO$_2$ price and GDP output metrics in 2050, and the key characteristics (modelling strength) of the individual models and analyses. As this project is *not* a focused model comparison based on an integrated assumption set (and also to make the table more legible), additional metrics (e.g. energy consumption) are instead noted in the model summaries below and discussed in detail in the subsequent papers.

3.1. Global models

Barker et al. use the global dynamic macro-econometric E3MG model and explore the role of targeted investment and induced technological change in achieving LCS. The E3MG model has a number of features that differentiate it from computable general equilibrium (CGE) models: it does not assume the global economy is in equilibrium or at full employment; it allows for varying returns to scale; it allows for different behaviours for each industrial sector and region; it incorporates endogenous technological learning; and it allows for revenue recycling and explicit tracking of investment flows by region and sector. It is linked to a technologically detailed bottom-up energy model which includes non-linear learning processes (including threshold effects) for energy technologies. The model's emissions targets include all major GHGs. The model baseline is based on that developed for the European Commission-funded 'Adaptation and Mitigation Strategies supporting European Climate Policy' (ADAM) project. A *Carbon price* run is modelled, as is the *Carbon-plus* run for a global 50% emission reduction. Further sensitivities investigate support policies for key sectors based on revenue raised, focusing on electricity, transport (plug-in electric vehicles), industry, dwellings (efficiency), and finally accelerating the rise in the carbon price to $100/tCO$_2$ by 2030 (remaining at that value to 2050), in order to realize the 50% reduction target under *Carbon-plus*.

The *Carbon price* scenario alone delivers substantial CO$_2$ reductions (15% below year 2000 levels, compared with a rise of 105% in the *Base case*), but with a global GDP loss of –1.9%. The largest emission reductions (but still highest growth rates) in CO$_2$ emissions are from developing countries, while the costs fall disproportionally on heavy coal-using countries. Under a *Carbon-plus* scenario, global GDP now rises by +1.1%. This is due to worldwide induced extra investments and/or R&D in alternative energy technologies as a consequence of both higher real carbon prices and the additional mitigation measures. Because the model does not assume that the economy is in equilibrium, these higher investments are associated with higher quality and innovatory products (i.e. a switch to low-carbon technologies), leading to greater exports, which in turn are effectively matched by an increase in supply that is realized in the model through economies of scale, learning curves and higher productivity. Large and fast-growing developing economies (India, China, Russia) benefit most from targeted policies. Of these policies, measures to promote more rapid innovation and deployment of low-carbon electricity and transport technologies are key to achieving additional emissions reductions, whilst improving overall GDP.

The authors recognize that their finding of global GDP rises, due to investment effects under a low-carbon scenario through recycling of revenues, contrasts with those of most macroeconomic models, most of which use a CGE framework. They argue that their approach provides a more realistic set of assumptions. This would imply that significant economic as well as environmental benefits would arise from the efficient and targeted use of carbon tax revenues to promote innovation and deployment of low-carbon technologies. For example, in the transport sector, they find that significant additional carbon reductions come from measures to accelerate diffusion of electric plug-in vehicles through technology-based agreements, alongside the rising carbon price. They argue that a long-term framework comprising expectations of a rising carbon price and agreements to stimulate innovation and diffusion of low-carbon technologies could help to stimulate such large-scale investments in technologies and infrastructure to overcome the lock-in of the current system.

Akimoto et al. use the DNE21+ global (77 regions) energy systems model, with a detailed energy supply and demand technology treatment. The *Base case* utilizes future population and GDP forecasts based on the IPCC SRES B2, together with current climate policies and plans for attainment of the Kyoto target. The *Carbon price* case assumes the long-term (2013–2050) global carbon price of $10/tCO$_2$ rising to $100/tCO$_2$. Two *Carbon-plus* cases were run: a 'Vision 50/50' case assuming that global CO$_2$ emissions in 2050 are reduced to half of those in 2004; and a 'Sectoral approach toward 50/50' case, where specific intensity targets by sector were set, based on the results of the 'Vision 50/50' case. Here, intensity targets are established for power sectors, other energy conversions, iron/steel, cement, aluminium, paper/pulp and chemical industries, vehicles (car, bus and truck), and appliances (e.g. TV, air conditioner, refrigerator). CO$_2$ reductions are smaller in the 'Sectoral approach toward 50/50' case, particularly in other industrial sectors in which the intensity targets were not assumed, transportation sectors, and residential and commercial sectors.

In terms of global net CO$_2$ emissions from fossil fuel combustion, from a *Base case* 2050 level of 58.0 GtCO$_2$, the *Carbon price* case produces 21.7 GtCO$_2$, the *Vision 50/50* case produces 13.0 GtCO$_2$ and the *Sectoral approach* case produces 22.5 GtCO$_2$. In terms of *averaged* costs of CO$_2$ emission reduction ($/tCO$_2$) in 2050, the *Carbon price* case gives a global price of $31.8/tCO$_2$, the *Vision 50/50* gives $49.6/tCO$_2$ and the *Sectoral approach* gives $45.4/tCO$_2$.

In the *Sectoral approach* case, large emission reductions come from efficiency and conservation in many sectors, together with power sector reductions from nuclear power, and fuel switching from coal to natural gas in power sectors up to 2020. After 2030, CO_2 storage contributes to large emission reductions, with 17.3 $GtCO_2$ of annual sequestration by 2050. Renewable power, together with innovative technologies in many sectors (e.g. advanced nuclear power, direct reduced iron and steel production using hydrogen accompanying CCS, plug-in hybrid and fuel-cell cars), are also required to contribute to deep emission reductions by 2050.

The authors argue that the proposed sectoral-based approach provides a more politically realistic strategy for achieving deep emission cuts. However, they recognize that in order to achieve the sectoral targets, strong and early R&D actions to accelerate the cost reductions in CO_2 emission reduction technologies are necessary, as well as actions to promote the public acceptability of widespread deployment of the full range of low-carbon technologies. Hence, integration is needed of a publicly accepted LCS vision and a sectoral-based framework to induce internationally strong R&D and deployment efforts.

Remme et al. utilize the global ETSAP-TIAM model. This multi-region, technology detailed, bottom-up model includes endogenous energy prices and emissions trading. The *Base case* is linked to the IPCC B2 scenario. Together with the $100/tCO_2$ *Carbon price* case, the *Carbon-plus* case of a global 50% emission reduction is supplemented with a sensitivity analysis on nuclear penetration.

In terms of CO_2 emissions, the *Base case* sees a doubling by 2050 (to 48 $GtCO_2$), driven by coal use in the power sector, especially in developing countries. The *Carbon price* case gives a 16% reduction from 2000 (60% reduction from base 2050 levels), including coal CCS capture (coal electric and H_2 production) at 10.5 $GtCO_2$. The *Carbon-plus* case delivers the 50% reduction from 2000 (72% reduction from base 2050 levels), with similar CCS levels, but now, after 2040, by the construction of natural gas plants with CO_2 capture and coal-fired oxy-fuel capture plants. In this supply-side optimization approach, relatively modest final energy reductions (11% in *Cprice*; 21% in *Cplus*) are overshadowed by power-sector reductions supplemented by transport hybrids, CHP and heat pumps.

In terms of abatement split between regions, developing countries realize the greater percentage reductions from base case trends but still constitute the majority of emissions in 2050. Economic impacts in 2050 in terms of additional energy system costs are $641bn (0.5% of global GDP in 2050) in the *Carbon price* case, and $1,633bn (1.3% of global GDP in 2050) in the *Carbon-plus* case. CO_2 marginal abatement costs are $100/tCO_2$ (average abatement costs $21/tCO_2$) in the *Carbon price* case, and $330/tCO_2$ (average $44/tCO_2$) in the *Carbon-plus* case. The nuclear sensitivity sees a similar pattern and costs, but with major substitutions between nuclear and CCS as power-sector abatement options.

Policy implications from a high fossil energy growth base case are that, for major CO_2 reductions, the power sector and CCS are key technologies. Hence, regulatory barriers to CCS deployment need to be addressed. This is complemented by renewable electricity, CHP, industrial efficiency, bio-ethanol transport, hydrogen (from coal with CCS) transport at higher carbon prices. However, in addition to the carbon price, a range of other policy measures are needed to stimulate their deployment, including research, demonstration, deployment programmes, information campaigns and standards, and investment incentives. Finally, burden sharing and emission trading are crucial, as in 2050 the majority of reductions are from developing countries, but these countries still have the majority of emissions.

The global model scenario runs are completed with *Edmonds et al.*, who utilize the ObjECTS MiniCAM model – a long-run, partial-equilibrium model (energy and land use) including numerous energy supply technologies, agriculture and land-use models, together with a reduced-form climate model. The two core runs described in this paper focus on a *Base case* with no climate policy, and stabilization (*Carbon-plus* case) of radiative forcing at 3.4 W/m², relative to a pre-industrial state. The authors argue that this is equivalent to stabilizing the concentration of CO_2 at approximately 450 parts per million (ppm), stabilizing the concentration of CH_4 at approximately 1.4 ppm, and stabilizing the concentration of N_2O at 0.36 ppm. In terms of CO_2 emissions, the *Base case* results in a global doubling by 2050 (to around 51 $GtCO_2$), driven by developing countries. The USA realizes a smaller increase from 5.9 $GtCO_2$ in 2000 to 7.0 $GtCO_2$ in 2050. The stabilization (or *Carbon-plus*) case equates to a 50% global reduction from 2010 levels, with much larger (58%) reductions in the USA.

The *Carbon-plus* case entails the price of carbon rising exponentially until mid-century where it reaches approximately \$136/tCO_2, at which point the concentration of CO_2 approaches its steady-state value. In terms of energy use, in 2050, US primary energy consumption is only about 20% smaller than in the reference scenario, although as an example the building sector emissions are driven down by more than 65%. This is driven by supply-side decarbonization, with by 2050 almost half of all primary energy provided by non-fossil energy forms; largely nuclear, CCS, solar, wind and biomass. In particular, nuclear energy doubles by 2040 and triples by 2070, while by 2050 all coal use employs CCS and by 2070 almost all fossil fuel power generation employs CCS technology.

The authors assume that global emissions reductions are achieved in an economically efficient manner across the world. Because limiting radiative forcing to 3.4 W/m² implies a limit in cumulative emissions, delayed participation on the part of any region means that the emissions mitigation that occurs in this idealized control regime must be made up by larger and less efficient emissions reductions in other regions.

3.2. National models

Bataille et al. use the CIMS hybrid technology simulation model: an integrated, energy–economy equilibrium model. CIMS contains bottom-up representation of 2,800 technologies competing to meet the demand for hundreds of final and intermediate goods and services, covering 15 energy and industrial sectors, and provides top-down simulation of the interaction of energy supply/demand and the macroeconomic performance of any economy, including trade effects.

The focus of this analysis was a comparison of Canada (a slow-growing developed economy) and China (a fast-growing developing economy), using national CIMS-Canada and CIMS-China models. In addition to *Base case* and *Carbon price* (to \$100/tCO_2 by 2050) the Canada *Carbon-plus* run was a 50% reduction in CO_2 emissions, while the China *Carbon-plus* run was stabilization at 2010 CO_2 emission levels.

For Canada, *Carbon-plus* is achieved through a combination of emissions pricing (doubling to \$200/tCO_2 by 2050) and regulatory measures in the transport and buildings sectors. For China, carbon stabilization to 2010 emissions levels requires a \$100/tCO_2 tax plus the inclusion of complementary regulations, namely accelerated closure of small power plants in the electricity industry, subsidies for renewables, accelerated decommissioning of inefficient heavy industrial plants, vehicle efficiency standards, and voluntary initiatives/public environmental campaign.

However, when modelling trade impacts, developed countries plus China could together achieve a 50% reduction in GHG emissions with a carbon price of \$175/tCO_2 with permits flowing *from* the developed countries *to* China. This is because of China's demand for energy to drive sustained economic growth. The policy implication is that, while all countries undertake major emissions reductions, in the early years the developing world provides greater opportunities for lower-cost emissions reduction, followed in later years by the developed countries. This dynamic relationship argues for flexible policy instruments, such as emissions cap-and-trade systems, which can adapt as circumstances change. In the long run, developed countries will probably be required to subsidize low-GHG development initiatives in the developing world (e.g. through knowledge transfer and the voluntary assumption of relatively tighter emissions caps).

Fujino et al. illustrate the feasibility of very deep (–70%) long-term reductions in Japan. Two scenarios are analysed using a linked set of models: the Population and Household Model (PHM), the Building Dynamics Model (BDM), the Transportation Demand Model (TDM), and the Computable General Equilibrium (CGE) model, for the Japanese economy.

The two scenarios to achieve 70% reduction in CO_2 emissions from Japanese economy by 2050 are:

- *Scenario A*: an active, quick-changing, and technology-oriented society, with an annual growth rate of per capita GDP of 2%.
- *Scenario B*: a calmer, slower, and nature-oriented society, with annual growth rate of per capita GDP of 1%.

In both scenarios, expected future innovations will lead to reductions in energy demand by 40–45% from the 2000 level while maintaining GDP growth and improving service demands. In addition, decarbonization of energy supply boosts CO_2 reductions to 70% below 2000 levels. Strategies for realizing LCS include three key elements: demand reduction through structural transformations to reduce energy service demands; development and deployment of energy-efficient technologies; and decarbonization of energy in the supply-side.

Scenario A shows that key decarbonization routes are energy efficiency options in demand-side, such as implementation of energy-efficient appliances in the industrial, residential, commercial and transportation sectors, and fuel-switching options from conventional energy sources to low-carbon energy sources, such as nuclear power and hydrogen. The macroeconomic cost is 0.83–0.90% of GDP in 2050. *Scenario B* shows that the use of low-carbon energy, such as biomass and solar energy, in demand-side, would result in drastic reductions of CO_2 emissions. The macroeconomic cost is 0.96–1.06% of GDP in 2050. Although CO_2 reductions vary by sector, both scenarios share many technology options:

- No-regret investments, which reduce energy costs and are profitable.
- R&D activities for such technologies that yield desirable outcomes for society.
- The technology options that take long periods of time for implementing, such as hydrogen, nuclear power and renewable-based energy systems, require early and well-planned strategies with consideration of uncertainties.

In order to achieve LCS by 2050 without missing opportunities for various investments on capital formation and technology development, it is necessary to set up the national goals (i.e. the vision of LCS, rather than target rates of reduction) at an early stage, establish the abatement schemes, and advance a societal change in behaviour that internalizes the negative externalities of CO_2 emissions. In this process, these goals stimulate and accelerate social and technological innovations that help in getting an advantage in international competition in the future low-carbon world.

Strachan et al. utilize a hybrid technology optimization model linked to a neoclassical growth model; the UK MARKAL-Macro (M-M) model. A core strength of a national energy model such as UK M-M is its detailed depiction of physical, economic and policy aspects of a country-level energy system. However, UK M-M must incorporate the critical role of international drivers on the UK's costs and pathways to a future LCS. These drivers are conventionally explored through extensive sensitivity analysis, and include:

- technology costs
- fossil fuel resource prices
- supply of imported resources, such as biomass
- international aviation emissions
- trading mechanisms for international CO_2 emission reductions permits.

The paper undertakes a *Base case* (based on UK energy demand forecasts), the $10 rising to $100/t$CO_2$ *Carbon price* run and a *Carbon-plus* run where the UK leads on international mitigation efforts and achieves a domestic CO_2 emission reduction of 80% by 2050. In addition, five sensitivity runs are undertaken on *Carbon-plus*, based on the role of the five international drivers. These are modelled individually and then grouped into an 'Annex 1 consensus' or a 'Global consensus' as integrated scenarios that depend on the scope of long-term emissions mitigation policies.

In 2050, the *Carbon price* case shows an absolute UK emission reduction of around 50% with GDP losses of only 0.33%, but this rises significantly to 1.64% (or $83bn) in the *Carbon-plus* (*Annex 1 consensus*) case and 2.21% in the *Global consensus* scenario (both 80% UK CO_2 reductions in 2050). This corresponds to a carbon price range of $402/t$CO_2$ rising to $490/t$CO_2$. When considering the range of international drivers, for aggregate results, the inclusion of international aviation and the potential large-scale purchases of CO_2 emissions reduction are most important. However, when all countries are meeting CO_2 reduction targets under LCS strategies (*Global consensus*), the availability and hence cost reduction impact of international purchases is drastically reduced. All international drivers have major implications for the sectoral and technology portfolio distribution of decarbonization efforts (especially in the electricity sector).

Concluding with developing-country national models, *Shrestha et al.* utilize the Asia–Pacific Integrated Assessment Model (AIM)/Enduse framework as a bottom-up energy system model with considerable detail on energy supply, technology pathway and energy service demands for Thailand. They investigated a *Base case* (through 2050) and three *Carbon tax* scenarios:

- C10+ (*Carbon price*): US$10–100/t$CO_2$ exponentially rising from 2013 to 2050
- C75: constant US$75/t$CO_2$ through to 2050
- C100: constant US$100/t$CO_2$ through to 2050.

In the *Base case*, CO_2 emissions are projected to grow more than sevenfold, i.e. from 158 MtCO_2 in 2000 to 1,299 MtCO_2 by 2050. Among the *Carbon tax* scenarios, C100 results in the highest cumulative CO_2 emission reduction (16.4%) followed by C75 (11.5%) and C10+ (5.5%). With the introduction of increasing levels of a carbon tax, reduction in CO_2 emission is mainly achieved in the power sector through the use of carbon capture and storage (CCS) technology in coal and natural gas power plants, with additional use of CCGT gas plants and nuclear power. The role of renewable electricity generation remains small. If a modal shift was enabled for passenger transport from low-occupancy vehicles to Mass Rapid Transit Systems (MRTS) and railways (from 10% in

2015 to 30% in 2050), the economy-wide reduction in CO_2 emissions would increase from 5.5% to 10.1% in C10+, from 11.5% to 16.5% in C75, and from 16.4% to 19.2% in the C100 case.

In terms of policy implications, using a partial equilibrium model of a fast-growing developing-country energy system shows the difficulty in realizing deep absolute CO_2 reductions. The majority of the CO_2 abatement (over 76%) would take place in the power sector, with CCS being a key technology. The model finds that biomass and other renewable energy technologies would not play a significant role, and thus Thailand would continue to depend largely on imported energy. Lastly, given the heavy reliance on low-occupancy personal vehicles for passenger transport, the study also shows a very significant potential for CO_2 emission reduction through modal shift in transport system to electrified MRTS and railways.

Finally, *Shukla et al.* use an integrated soft-linked model framework with the AIM CGE model for GDP projections and the Indian MARKAL model providing detailed technology- and sector-level energy and emissions with end-use demand modelling and energy balance calculations. Scenarios run were the *Base case*, and a *Carbon tax* (CT) run (rising to $100/tCO_2$ in 2050 and corresponding here to a global 550 ppm CO_2e target), and a *Sustainability scenario* (SS), which achieves mitigation equal to that in the CT scenario by alternative measures. Like other developing countries, the base case illustrates India's huge growth projections through 2050 with GDP rising by +2,360%, and CO_2 emissions by +414%, leading to cumulative emissions (2005–2050) of 162.3 GtCO$_2$. This is despite an energy intensity decrease of –3.29% per annum.

The CT scenario in 2050 gives a GDP loss for India of –1.35%, with Indian CO_2 emissions dramatically reduced from the base case but still entailing a 50% increase on 2005 levels (with cumulative emissions of 99.7 GtCO$_2$). CO_2 emissions reductions come mainly from CCS (19.1 GtCO$_2$) and fuel-switching in the electricity sector (30.5 GtCO$_2$).

In the *Sustainability scenario*, GDP is not reduced relative to the base case, as final consumption of goods and services is not curtailed. Instead, a range of policies to reduce intermediate demand for products are enacted, including modal shifts in transport, urban planning, building design and material substitution. Although 2050 CO_2 emissions are higher, with a +171% increase, the cumulative emissions reductions are the same as the CT case at 99.7 GtCO$_2$. CO_2 emissions reductions come mainly from the range of measures to reduce intermediate demand for products (48.7 GtCO$_2$) and fuel-switching in the electricity sector (13.4 GtCO$_2$).

The key policy implication is that a sustainability scenario achieves the same cumulative emissions reduction as a carbon tax scenario, but at no cost to GDP and with significant co-benefits. These co-benefits include reductions in local air pollutants (SO_2 and NO_x) and an increase in capacity for adaptation. In addition, cooperation among stakeholders and reduced transaction costs realize a higher deployable potential of renewable resources. This alternate paradigm embeds the LCS transition and low-carbon choices within larger development issues.

4. Discussion and key policy implications

This LCS modelling project has illustrated a complex picture of a possible future transition to a low-carbon society. This international modelling exercise used a range of global or national energy models, focused on macroeconomic or technology-driven approaches (or more commonly a hybrid of the two). Individual modelling assessments of long-term LCS scenarios concentrated on key aspects of such transitions, and as such delivered a broader range of policy insights. This approach meant that precise model comparisons based on an integrated set of assumptions are not applicable,[6] with that method being excellently covered in earlier comparative exercises (EMF-23, IMCP).

However, common themes for policy makers do emerge, which are further substantiated in the individual modelling papers.

The models focused on core drivers, depending on their individual strengths. Predominant in these are technological change, international emissions trading, non-price (sustainable development) mechanisms, and behavioural change in energy service demands. The GDP annual losses (in 2050) of achieving LCS generally ranged from –0.35% to –1.35%, although these rose to –2.20%, depending on target stringency. Interestingly, one global econometric model (E3MG) and one integrated soft-linked (AIM and MARKAL) model found the change in GDP could be zero or even positive (i.e. net economic benefit) with judicious sustainable planning and/or targeted use of emission revenues to promote investment in innovation and deployment of low-carbon technologies. In terms of carbon price signals, the general range was between $100/tCO$_2$ and $330/tCO$_2$, again rising to $490/tCO$_2$ depending on target stringency.

In terms of policy implications, a fundamental finding is that LCS scenarios are technologically feasible given expected progress in low-carbon measures and the behavioural change required to adopt technologies and complement them with emissions reductions. However, they require clear and early target setting and incorporation of emissions targets in every facet of energy use and economic activity. The issue of the optimal timing of CO$_2$ emissions reduction is somewhat controversial. For this study, diagnostic model runs show that delaying setting emissions targets may result in final targets still being met. However, the *Carbon-plus* runs generally find the models' preferred solution is for some level of early action to stimulate the development and cost reduction in technologies, facilitate the turnover in energy infrastructures, and embody societal change in the demand for energy services.

The LCS scenarios will probably entail significant socio-economic changes at the company and individual levels. To realize these very major technological transitions requires sustained progress in technical and cost improvements in low-carbon technology options. Some of the comparison models (as discussed in the individual papers) emphasize either or both the endogenous role of returns from R&D and/or learning from production economies of scale. Other models that do not encompass endogenous technological change utilize sensitivity analysis to demonstrate the necessity of accelerated technological progress in a broad portfolio. In a portfolio of low-carbon options, carbon capture and storage (CCS) is found in many models to be a key technology. Highlighting CCS technologies as a key emission reduction option does not diminish the potential importance of nuclear and renewables. These large-scale zero-carbon electricity technologies are judged on their relative costs, energy system interactions, as well as (in some models) their political feasibility.

Developing countries, in particular, face an immense challenge to achieve LCS in light of their projected economic growth and energy use requirements. As such, international cooperation is required, notably in flexible burden sharing under international emissions trading regimes. In the long term, however, it is not clear which regions will have the lowest-cost emissions reductions, and hence an iterative strategy for interim targets and burden sharing (under a long-term target) appears most appropriate.

Notes

1. See www.ukerc.ac.uk/TheMeetingPlace/Activities/Activities2006/0612QuantEnergyScenariosLCS.aspx.
2. Notably on demands for energy services.
3. Including institutes previously less integrated into modelling networks.
4. For example, the model target may vary due to country-specific factors – e.g. Annex 1 vs. developing-country targets; the model target may also vary due to consideration (or not) of land-use CO$_2$ and non-CO$_2$ GHGs.

5. Note that a team from the Energy Research Institute, China, participated in the workshop discussions but did not contribute a final paper for inclusion in this supplement.
6. This would be further complicated due to alternative baselines (around standard IPCC or other national projections), and alternative definitions of burden sharing under a global LCS defined as a 50% reduction in GHG emissions by 2050.

References

Barrett, S., 2003, *Environment and Statecraft: The Strategy of Environmental Treaty Making*, Oxford University Press, Oxford, UK.

DEFRA, 2007, *Draft Climate Change Bill, March 2007*, Department for Environment, Food and Rural Affairs, London [available at www.official-documents.gov.uk/document/cm70/7040/7040.pdf].

Edenhofer, O., Lessmann, K., Kemfert, C., Grubb, M., Koehler, J., 2006, 'Endogenous technological change and the economics of atmospheric stabilization: synthesis report from the Innovation Modelling Comparison Project', *Energy Journal* Special Issue 1.

G8 Communiqué, 2007, *Chair's Summary*, G8 Heiligendamm Summit, 8 June 2007 [available at www.g-8.de/Webs/G8/EN/G8Summit/SummitDocuments/summit-documents.html].

IEA, 2006, *Energy Technology Perspectives: Scenarios and Strategies to 2050*, International Energy Agency, Paris.

IPCC, 2007a, *Synthesis Report of the IPCC Fourth Assessment Report*, Intergovernmental Panel on Climate Change, Cambridge University Press, Cambridge, UK.

IPCC, 2007b, *Climate Change 2007: Mitigation.* Contribution of Working Group III to the Fourth Assessment Report of the Intergovernmental Panel on Climate Change, Cambridge University Press, Cambridge, UK.

Nordhaus, W., 1994, *Managing the Global Commons: The Economics of Climate Change*, MIT Press, Cambridge, MA.

Pearce, D., 2003, 'The social cost of carbon and its policy implications', *Oxford Review of Economic Policy* 19(3), 362–384.

Skea, J., Nishioka, S., 2008, 'Policies and practices for a low-carbon society', *Climate Policy* 8, Supplement, 2008, S5–S16.

Stern, N., 2006, *The Economics of Climate Change: The Stern Review for HM Treasury, London*, Cambridge University Press, Cambridge, UK.

van Vuuren, D., Weyant, J., de la Chesnaye, F., 2006, 'Multi-gas scenarios to stabilize radiative forcing', *Energy Economics* 28, 102–120.

climate
policy

■ research article

Achieving the G8 50% target: modelling induced and accelerated technological change using the macro-econometric model E3MG

TERRY BARKER[1]*, S. SERBAN SCRIECIU[1], TIM FOXON[2]

[1] Cambridge Centre for Climate Change Mitigation Research (4CMR), Department of Land Economy, University of Cambridge, 19 Silver Street, Cambridge CB3 9EP, UK

[2] Sustainability Research Institute (SRI), School of Earth and Environment, University of Leeds, Leeds LS2 9JT, UK

This article assesses the feasibility of a 50% reduction in CO_2 emissions by 2050 using a large-scale Post Keynesian simulation model of the global energy–environment–economy system. The main policy to achieve the target is a carbon price rising to $100/t$CO_2$ by 2050, attained through auctioned CO_2 permits for the energy sector, and carbon taxes for the rest of the economy. This policy *induces* technological change. However, this price is insufficient, and global CO_2 would be only about 15% below 2000 levels by 2050. In order to achieve the target, additional policies have been modelled in a portfolio, with the auction and tax revenues partly recycled to support investment in low-GHG technologies in energy, manufacturing and transportation, and 'no-regrets' options for buildings. This direct support supplements the effects of the increases in carbon prices, so that the *accelerated* adoption of new technologies leads to lower unit costs. In addition the $100/t$CO_2$ price is reached earlier, by 2030, strengthening the price signal. In a low-carbon society, as modelled, GDP is slightly above the baseline as a consequence of more rapid development induced by more investment and increased technological change.

Keywords: carbon pricing; CO_2 reductions; G8 CO_2 50% target; global economy model; induced technological change; low-carbon society; scenario modelling

Ce papier analyse la faisabilité d'une réduction de 50% des émissions de CO_2 d'ici 2050 par le biais d'un modèle de simulation post keynésien de grande échelle du système mondial énergie-environnement-économie. La principale politique pour atteindre cet objectif est l'augmentation du prix du carbone à $100/t$CO_2$ d'ici 2050, obtenu par l'auctioning de permis de CO_2 pour le secteur de l'énergie et des taxes carbone dans les autres secteurs. Cette politique induit le changement technologique. Cependant, ce prix est insuffisant et globalement le CO_2 ne serait réduit en 2050 que de 15% au-dessous du niveau de 2000. Dans le but d'aboutir à cet objectif, des politiques supplémentaires ont été modélisées dans un portefeuille, en recyclant partiellement les revenus des ventes aux enchères et des taxes pour encourager l'investissement dans les technologies sobres en GES, dans l'énergie, la production et le transport et les options sans regret dans le bâtiment. Ce soutien direct s'ajoute aux effets des augmentations du prix du carbone, l'adoption accélérée de nouvelles technologies entraînant ainsi des prix unitaires inférieurs. De plus les valeurs de $100/t$CO_2$ sont obtenues plus tôt, d'ici 2030, renforçant ainsi le signal prix. Dans la société sobre en carbone, telle que le modèle le montre, le PIB est légèrement supérieur à la baseline en conséquence d'un développement induit plus rapide dû à deavantage d'investissements et un changement technologique accru.

Mots clés: changement technologique induit; fixation du prix du carbone; l'objectif de réduction de CO_2 de 50% du G8; modèle de l'économie mondiale; modélisation de scenarios; réductions de GES; société sobre en carbone

■ *Corresponding author. E-mail*: tsb1@cam.ac.uk

doi:10.3763/cpol.2007.0490 © 2008 Earthscan ISSN: 1469-3062 (print), 1752-7457 (online) www.climatepolicy.com

1. Introduction

This article contributes to the UK–Japan modelling comparison project by developing and analysing global low-carbon society (LCS) scenarios using a global macro-econometric model, E3MG. In common with the other contributions, we set out three scenarios: (i) a *base case* and (ii) a rising *carbon price* case, both of which include endogenous technological change;[1] and (iii) a *carbon-plus* case, which incorporates additional policy measures to accelerate the development and deployment of low-carbon technologies through regulation and fiscal incentives.

The emphasis in the results is on the modelling of a policy portfolio in the carbon-plus scenario. Most of the studies reviewed in The IPCC's Fourth Assessment Report (AR4) (IPCC, 2007) on induced technological change (e.g. Sijm, 2004; Edenhofer et al., 2006) consider how such change affects the costs associated with correcting the market failure of damaging GHG emissions. However, when the beneficial externalities from low-carbon innovation[2] are considered in the modelling, we can address a second market failure of insufficient innovation. This implies that at least two instruments should be included in any portfolio for policy choice (Clarke and Weyant, 2002, p.332; Fischer, 2003; Jaffe et al., 2005). Mitigation policies and technology policies can be implemented by a wide range of instruments (tools) and measures, and their effectiveness will depend on the mix and the implementation strategies. In other words, it is not necessarily a matter of pursuing mitigation and technology policies *per se*, but rather the scope of their adoption and the extent to which the interactions between them are adequately considered. Several studies have compared mitigation policies and technology policies, with the general finding that technology policies alone tend to have smaller impacts on emissions than mitigation policies do (Nordhaus, 2002; Fischer and Newell, 2004; Yang and Nordhaus, 2006). Here we combine the two.

We model several measures in our portfolio for the carbon-plus scenario by combining an escalating carbon price (via a global CO_2 emission trading scheme with auctioned permits for the energy sector and carbon taxes for the rest of the economy) with policies aimed at providing incentives to induce more rapid technological change across all energy users. We compare this carbon-plus scenario with the carbon price scenario without the extra incentives. Both scenarios and the base case allow for endogenous technological change.

Other researchers have also investigated such a combination.[3] In a short-run context, Masui et al. (2005) examined the effects of a carbon tax in Japan to meet the Kyoto target using the AIM (Asia–Pacific Integrated Model). By 2010, a carbon tax with lump-sum recycling would lead to an average GDP loss of 0.16% and a tax of US$115/$tCO_2$. However, a tax and subsidy regime, in which carbon tax revenues are used to subsidize CO_2 reduction investments, leads to an average GDP loss of 0.03% and a tax of US$9/$tCO_2$. In a long-run context, Weber et al. (2005), using a disequilibrium, calibrated global growth model, MADIAM, concluded that

> increasing the fraction of carbon taxes recycled into subsidizing investments in mitigation technologies not only reduces global warming, but also enhances economic growth by freeing business resources, which are then available for investments in human and physical capital (p. 321).

However MADIAM is a one-sector model, intended to explore ideas rather than to provide quantitative conclusions.[4]

The contribution that this article makes is to introduce a policy portfolio much more explicitly into a large-scale, long-term estimated global model, developing the portfolios of policies studied by Masui et al. (2005) for Japan to 2020, and at a one-sector, calibrated global level by Weber et al. (2005).

Our approach involves the use of econometric estimation to identify the effects of endogenous technological change on exports and energy demand and to embed these in a large Post Keynesian non-linear simulation model with a dynamic structure, which is both sector- and region-specific. A treatment of substitution between fossil fuel and non-fossil fuel technologies is included, accounting for non-linearities resulting from investment in new technology, learning-by-doing and innovation. The large-scale econometric model, E3MG (Energy–Environment–Economy Model at the Global level) then allows policy measures for induced technological change (ITC) to be modelled. The model is then used to undertake a detailed analysis of the macroeconomic implications of the *carbon price* and *carbon-plus* scenarios, and thus is able to assess the energy and economic policy consequences of moving towards a 50% reduction in global CO_2 emissions by 2050.

2. Description of the model

The hybrid model, E3MG, is a 20-region, structural, annual, dynamic, econometric simulation model based on data covering the period 1970–2001, and projected forward to 2050 or 2100 (Barker et al., 2006). The emphasis in the modelling is on two sets of estimated equations included in the model: aggregate energy demand by 19 fuel users and 20 world regions; and exports of goods and services by 41 industries and 20 regions. Each sector in each region is assumed to follow a different pattern of behaviour within an overall theoretical structure, implying that the representative agent assumption is invalid.[5] The model has now been extended to cover the basket of six greenhouse gases.

The E3MG industrial and energy/emissions database[6] covering the years 1970–2002 is drawn from OECD, IEA, GTAP, RIVM, and other national and international sources, processed to provide comprehensive and consistent time-series of varying quality and reliability across regions and sectors. It contains information about historic changes by region and sector in emissions, energy use, energy prices and taxes, input–output coefficients, and industries' output, trade, investment and employment. This is supplemented by data on macroeconomic behaviour from the IMF and the World Bank. However, the data are far from comprehensive, especially for developing countries. Therefore there are large uncertainties in the estimates for some regions and variables, which must be taken into account in the econometric estimation and in interpreting the results. Furthermore, the parameters based on 33-year historical data may not be appropriate for solutions that cover a period of 100 years. However, the E3MG modelling approach assumes that understanding the future is best done by first understanding the past; hence the econometric basis of the model.

E3MG represents a novel approach to the modelling of technological change and economic growth in the literature on the costs of climate stabilization. It is based upon a Post Keynesian economic view of the long-term picture, adopting the 'history' approach[7] of cumulative causation and demand-led growth (Kaldor, 1957, 1972, 1985; Setterfield, 2002), focusing on gross investment (Scott, 1989) and trade (McCombie and Thirlwall, 1994, 2004), and incorporating technological progress in gross investment enhanced by R&D expenditures. Growth in this approach is dependent on waves of investment and as a macroeconomic phenomenon arising out of increasing returns (Young, 1928), which lead to technological change and diffusion. Other Post Keynesian features of the model include: varying returns to scale (which are derived from estimation), non-equilibrium, not assuming full employment, varying degrees of competition, the feature that industries act as social groups and not as a group of individual firms (i.e. no optimization is assumed but bounded rationality is implied), and the grouping of countries and regions based on political criteria.

The model has been developed to include the bottom-up energy technology model (ETM) (Anderson and Winne, 2004) within a top-down macroeconomic model. Thus, like the studies of Nakicenovic and Riahi (2003) and McFarland et al. (2004), our modelling approach avoids the typical optimistic bias often attributed to a bottom-up engineering approach, and the unduly pessimistic bias of typical macroeconomic approaches. The advantages of using this combined approach have been reviewed by Grubb et al. (2002). E3MG thus incorporates endogenous technological change in three ways:

1. The sectoral energy and export demand equations include indicators of technological progress in the form of accumulated investment and R&D.
2. The ETM incorporates learning-by-doing through regional investment in energy generation technologies that reduce costs depending on global-scale economies.
3. Extra investment in new technologies, in relation to baseline investment, induces further output and therefore investment, trade, income, consumption and output in the rest of the world economy through a Keynesian multiplier effect.

These mechanisms describe the key features of accounting for endogenous technological change in E3MG. However, further changes can be induced by policy; hence the term *induced* technological change. For example, feed-in tariffs for renewables (as used in Germany) will alter relative prices such that investments in renewable technologies are stimulated and, depending on their learning curve characteristics (and Keynesian multiplier effects at the macro level), they will lead to higher adoption rates. The effects of technological change modelled in this way may turn out to be sufficiently large in a closed global model to account for a substantial proportion of the long-run growth of the system.

The bottom-up ETM component of E3MG is an annual, dynamic technology model developed from the approach described by Anderson and Winne (2004). It is based on the concept of a price effect on the elasticity of substitution between competing technologies. Existing economic models usually assume constant elasticities of substitution between competing technologies. Although the original ETM is not specifically regional and is not estimated by formal econometric techniques, it does model, in a simplified way, the switch from carbon energy sources to non-carbon energy sources over time. It is designed to account for the fact that a large array of non-carbon options is emerging, although their costs are generally high relative to those of fossil fuels. However, costs are declining relatively with innovation, investment and learning-by-doing. The process of substitution is also argued to be highly non-linear, involving threshold effects. The ETM models the process of substitution, allowing for non-carbon energy sources to meet a larger part of global energy demand as the price of these sources decreases with investment, learning-by-doing and innovation. The model considers 26 separate energy supply technologies, of which 19 are carbon-neutral. The estimates of learning rates come from a review of the literature (McDonald and Schrattenholzer, 2001), which found that new energy technologies, such as wind and solar PV, typically show learning rates of 10–30% (i.e. a 10–30% reduction in unit costs for a doubling of installed capacity), whereas more mature technologies show much lower learning rates, e.g. 6% for nuclear power.

Thus, compared to the existing modelling literature targeted at achieving the same goals, we argue that the advantages of the E3MG model lie in three main areas. First, the detailed nature of the model allows the representation of fairly complex scenarios, especially those that are differentiated according to sector and country or region. Similarly, the impact of any policy measure can be represented in a detailed way with the disaggregation of energy and environment industries for which the energy–environment–economy interactions are central. Second, the econometric grounding of the model makes it better able to represent behaviour (in contrast to mainstream computable general equilibrium models that are based on a single year's dataset and make huge assumptions on crucial parameter values). And, third, an interaction (two-way feedback) between the economy, energy demand/supply and environmental emissions is an undoubted advantage over other models, which may either ignore the interaction completely or can only assume a one-way causation.

3. Scenarios

The specifications of the parameters and assumptions for the base case, carbon price and carbon-plus scenarios are described in the following sections.

3.1. Base case scenario

The common ADAM baseline has been taken as a starting point.[8] The development of the baseline has entailed the use of a series of projections of macroeconomic variables to which E3MG is being calibrated. The most important of these refer to population, oil prices and GDP. Baseline population projections follow the UN medium-fertility variant projections up to 2050, which have been reorganized according to E3MG's regional structure and six population groups (male/female for ages <15, working age, retired). Global oil prices have been updated and baseline projections have been taken from the POLES model.[9] Baseline (endogenous but calibrated) GDP projections are partially derived from a global vector auto-regressive (GVAR) model,[10] which in turn employs the POLES oil prices as conditioning assumptions. GVAR estimates and projects dynamics of the world economic system that link exchange rate, interest rate, oil price and GDP variables across markets and countries to account for increased economic and financial integration. However, because GVAR has short-term features (no technological change, no population growth), the projected growth rates from GVAR have been combined with those used in the European Commission's WETO-H2 baseline scenario over the medium term (2010ri–2015), with the latter being solely adopted afterwards over the longer term (2015–2050).[11] In addition, as GVAR only has an aggregate of the Euro area, some soft links with E3ME have been developed to obtain the GDP projections for E3MG individual (and aggregate) European countries (regions). Other variables to which E3MG is calibrated in the base case include: CO_2 emissions from land-use change up to 2050 sourced from the TIMER model;[12] primary energy use in terms of growth rates from the POLES model; and other (regional) energy prices, world commodity prices, global inflation, short- and long-term interest rates, and exchange rate projections over the period 2000–2050, based on information received from Cambridge Econometrics.[13]

In addition, a sophisticated method has been developed to calibrate the baseline GDP components of E3MG over the 50-year projection period. The growth rates of the totals for gross output, net output, private consumption, government consumption, investment, exports and imports have been projected based on past trends (i.e. econometric work on time-series) and interlinkages between the respective variables, their lags, the growth in global GDP, the growth in country/regional GDP, and time and country dummies. This panel data analysis has been used to make projections of several key macroeconomic variables conditional on the regional and global GDP projections being adopted by the ADAM project. These macro variable totals for each region have then been matched with the model's projections at a sectoral level, maintaining adding-up constraints. The projections have been made subject to two further constraints: first the growth rates of global exports and imports have been matched at a 41-sectoral level, assuming that any imbalances remain constant at the levels in the historical data; and, second, the GDP identity has been imposed, such that GDP (expenditure basis) equals consumption plus investment plus exports minus imports, at a regional level. These methods are intended to ensure that the structural projections of E3MG in the base case scenario will reproduce, more or less, the changes in structure shown in the data period (1970–2002), with the Social Accounting Matrix identities maintained.

Although model projections are based on historical data, it is worth noting that any expected major shifts in industrial structures or fundamental relations are, at best, only partially captured. These are uncertainties inherent in all modelling and policy making; e.g. the emergence of completely new technologies is difficult to predict. However, our approach is that in order to think about the future we need to at least understand past observations; hence the historical approach in deriving the significance of relationships between variables in our model.

3.2. Carbon price scenario

This scenario has a global real (year 2000) carbon price of $2.5/tCO$_2$ in 2011, rising by $2.5 per year to $100/tCO$_2$ in 2050. There are no other measures inducing technological change other than the rise in carbon prices applying at a global level. However the carbon price is reached through a mix of an auctioned permit scheme for the energy sector (power generation, other energy and the energy-intensive industries such as steel, cement and chemicals) and carbon taxes for the rest of the economy, with all revenues recycled at the country/region levels. The revenue recycling is assumed to be done through reductions in indirect taxes so that the overall inflation rate is unaffected. In effect, the assumption is that of global environmental fiscal reform in which the overall macroeconomic balance is maintained by the fiscal and monetary authorities, and the gradually increasing real carbon prices are offset in their effects on inflation by gradual reductions in all other prices. Though we fully acknowledge that a global environmental fiscal reform is a strong assumption and may be difficult to achieve, our aim is to illustrate a policy of implementing emission reduction measures that covers at least all the major sectors and all major regions.

3.3. Carbon-plus scenario

Carbon-plus (Cplus) is a scenario for achieving the LCS target of a 50% global reduction in CO$_2$ emissions by 2050. This target is in line with that agreed by global leaders at the 2007 G8 summit, although we have taken it as referring to 50% of 2000 emissions. This is in line with a trajectory for achieving atmospheric stabilization of GHG concentrations at a level of approximately 500 ppm CO$_2$e.

This is driven by a set of incentives to encourage the deployment of low-carbon technologies, in addition to the carbon price. These are designed to give rise to economies of scale and economies of specialization in the deployment of these technologies. Learning rates for different technologies in different regions (leading to a reduction in costs with their cumulative deployment) correspond to measures to stimulate the deployment of those technologies. These measures are assumed to be funded by recycling of revenues raised from the rising carbon price.

For the purpose of achieving the LCS target within our modelling structure, we have constructed the Cplus scenario from a combination of intermediate scenarios which apply five types of stylized policy measures, each intended to contribute to extra investments in low-carbon technologies. The first (ADDelect) refers to using part of the revenue obtained from auctioning carbon permits to subsidise low-carbon electricity generation technologies. The subsidy (in terms of $ per kWh) is evenly spread across new technologies, i.e. renewables and CCS (excluding nuclear and hydro). The share of revenue recycled that is used for subsidizing the new technologies starts at a level of 40% from 2011 to 2030, dropping to 20% by 2040 and to 0% by 2050. The second measure (ADDtrans) implemented in the model brings in the accelerated diffusion of electric plug-in vehicles through technological agreements allowing for the decarbonizing of the transport sector. The third (ADDind) returns the revenue raised through auctioning to energy-intensive industries in order to incentivize the conversion to low-carbon production methods. The fourth (ADDdwell) allocates 20% of the revenue recycled from the carbon tax to investments in energy efficiency in households, by providing incentives for improving the energy efficiency of domestic dwellings and appliances, and for introducing new ones such as low-emission dwellings and solar appliances. Finally, the fifth measure accelerates the increase in carbon prices to reach $100/tCO$_2$ by 2030 instead of 2050. It is important to note that the Cplus scenario is constructed in a cumulative manner, i.e. each of the component scenarios includes additional measures to the previous one. For the purpose of clarity of exposition, we employ the following notations in our discussion of results:

ADDelect: Cprice plus additional incentives for electricity technologies
ADDtrans: as above (i.e. ADDelect) plus additional incentives for transport technologies
ADDind: as above plus additional incentives for industrial end-use technologies
ADDdwell: as above plus additional incentives for efficiency of dwellings
Cplus: as above plus accelerated permit price increase to $100/tCO$_2$ by 2030.

The mitigation and technology policies mentioned above apply across all low-carbon technologies that are being targeted. The principle behind this approach is that innovation within a particular subset of low-carbon technologies will occur largely depending on barriers, relative prices and the technical characteristics and adoption rates of each technology that is being supported.

4. Model outputs

E3MG outputs a large array of variables across the projection period. For the purpose of the UK–Japan modelling comparison exercise, we provide a summary of results only for CO$_2$ emissions from all anthropogenic sources, GDP and final energy consumption across major economies. Results are reported on a ten-yearly basis over the period 2000–2050 for all three scenarios: base case, carbon price and carbon-plus.

4.1. Carbon price scenario

A first result is that the implementation of the mitigation scenario that involves only a gradual increase in real (2000 US$) carbon prices to reach $100/tCO$_2$ (equivalent to $367/tC) by 2050 only reduces global CO$_2$ emissions to 15% below 2000 levels and leads to lower overall world economic growth (see Figure 1). Global GDP in 2050 is projected to achieve a level, in the carbon price scenario, of $101.3 trillion, which is below (by around 1.9%) the corresponding baseline level of $103 trillion[14] (starting from a level of $30.1 trillion in 2000). The lower growth is not due to any potential increase in inflation (which may lead to increases in interest rates and lower growth), but to the sharp reduction in fossil fuel investments after 2010 when price expectations change following the introduction of an escalating carbon price. These fossil fuel investments become unprofitable, especially in China, but at the same time the expected growth in energy and electricity demand is reduced by the rising carbon price, so that the fossil fuel investments in the baseline are not entirely replaced by low-carbon investments in the scenario. Although alternative clean technologies might display a perceived additional cost premium associated with a higher than 'normal' economic and technical risk, we argue that this is mitigated in the long run by learning-by-doing and the

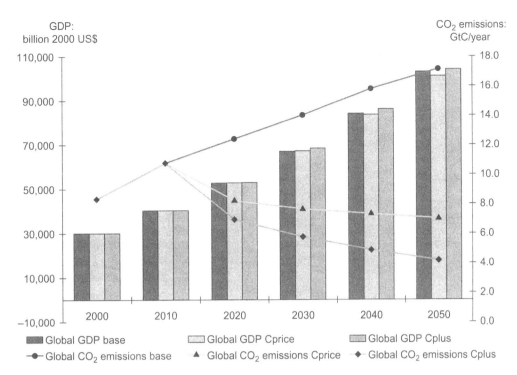

FIGURE 1 Total CO_2 emissions and GDP at the global level: base case versus carbon price scenario versus carbon-plus scenario, 2000–2050.

Source: E3MG2.3 SP7 baseline, LCS and LCP scenarios.

diffusion and widespread adoption of low-carbon technologies. In addition, portfolio analysis of risk suggests that the optimum mix of technologies favours renewables, because of the volatility of fossil fuel prices (Awerbuch, 2006). In consequence, the overall level of investment is reduced, leading to slower GDP growth. In the longer term (to 2100), the process works the other way, and low-carbon investments in the scenario are higher than fossil fuel investments in the base case.[15]

The emissions pathways associated with the base case and the carbon price scenario (as well as for the carbon-plus scenario discussed later in this section) are displayed in Figure 1, alongside GDP projections. CO_2 emissions from all anthropogenic sources[16] in the carbon price scenario decline from 10.8 GtC/year in 2010 (when carbon permit and tax schemes begin to apply) to 7 GtC/year in 2050 (equivalent to a reduction of approximately 60% relative to the projected 2050 baseline levels). This is associated with only a 16% reduction compared with 2000 levels (CO_2 emissions increase in the baseline from 8.3 to 17.1 GtC/year over the period 2000–2050).

The corresponding changes in CO_2 emissions and GDP effects across major world economies associated with the base case and the carbon price scenarios are shown in Table 1.[17] As expected, CO_2 emissions decline across all economies in the carbon price scenario. The largest cuts in CO_2 emissions by 2050 are expected to occur in China and the USA (by 83% and 60%, respectively, relative to baseline projections). This is again due to the heavy reliance of these countries on existing and planned high-CO_2-emitting coal-fired plants, which are argued to decline with the introduction of high real carbon prices and the corresponding increase in the costs of producing energy sourced from fossil fuels, in particular (conventional) coal.

TABLE 1 Total CO_2 emissions and GDP effects across major world economies: base case versus carbon price (Cprice) scenario, 2000–2050[21]

CO_2 emissions of all anthropogenic sources	2000	2010	2020	2030	2040	2050
EU-25 base (GtC/year)	1.21	1.18	1.19	1.23	1.31	1.31
EU-25 Cprice (% change from base)	0.0	0.0	−20.2	−29.5	−36.4	−39.2
USA base (GtC/year)	1.72	1.84	2.04	2.25	2.45	2.70
USA Cprice (% change from base)	0.0	0.0	−39.2	−49.0	−56.2	−60.5
Japan base (GtC/year)	0.37	0.39	0.41	0.45	0.41	0.45
Japan Cprice (% change from base)	0.0	0.0	−11.5	−24.5	−22.4	−32.4
Russia base (GtC/year)	0.48	0.54	0.49	0.51	0.61	0.60
Russia Cprice (% change from base)	0.0	0.0	−16.5	−35.9	−41.8	−37.3
China base (GtC/year)	1.39	2.34	3.07	3.95	5.07	5.08
China Cprice (% change from base)	0.0	0.0	−51.4	−71.1	−75.4	−82.7
India base (GtC/year)	0.34	0.53	0.81	0.85	0.90	1.04
India Cprice (% change from base)	0.0	0.0	−43.6	−37.8	−44.9	−47.7
Rest of World base (GtC/year)	2.81	3.96	4.40	4.77	5.06	5.97
Rest of World Cprice (% change from base)	0.0	0.0	−24.8	−31.0	−40.7	−49.1
World total base (GtC/year)	8.32	10.78	12.40	14.02	15.80	17.14
World total Cprice (% change from base)	0.0	0.0	−33.8	−45.5	−53.7	−59.1
GDP	2000	2010	2020	2030	2040	2050
EU-25 base ($ billion)	8,100	10,002	12,272	14,438	16,950	19,575
EU-25 Cprice (% change from base)	0.0	0.0	0.1	0.4	−0.2	−1.6
USA base ($ billion)	9,810	12,840	16,127	19,842	23,978	27,956
USA Cprice (% change from base)	0.0	0.0	0.1	0.3	0.4	0.3
Japan base ($ billion)	4,765	5,729	6,797	8,051	9,710	11,392
Japan Cprice (% change from base)	0.0	0.0	0.0	0.3	−0.5	−1.7
Russia base ($ billion)	260	437	591	735	878	1,036
Russia Cprice (% change from base)	0.0	0.0	2.7	3.7	3.8	4.4
China base ($ billion)	1,243	2,874	5,042	7,416	10,577	14,339
China Cprice (% change from base)	0.0	0.0	−0.2	−0.1	−5.2	−10.1
India base ($ billion)	457	891	1,507	2,293	3,319	4,633
India Cprice (% change from base)	0.0	0.0	0.6	2.0	1.7	2.2
Rest of World base ($ billion)	5,496	7,610	10,560	14,236	18,751	24,107
Rest of World Cprice (% change from base)	0.0	0.0	0.1	0.0	−0.4	−0.7
World total base ($ billion)	30,131	40,383	52,896	67,011	84,163	103,038
World total Cprice (% change from base)	0.0	0.0	0.1	0.3	−0.6	−1.9

Source: E3MG2.3 SP7 baseline and LCS scenarios.

Setting carbon prices to reach $100/tCO$_2$ in 2050 results in a positive impact on economic output (relative to the baseline) for some countries/regions, and a negative real GDP effect for others over the projected period. The latter is particularly the case for China, for which the projected GDP for 2050 is around 10% lower in the carbon price scenario relative to the baseline. This may be attributed to the fact that, in the baseline, China heavily relies on energy use from coal to meet the accelerating energy demand deriving from high growth rates, leading to substantial planned investments in coal-fired plants. As explained above, with the introduction of high carbon prices there is a collapse in the plans for these investments which, combined with reduced energy demand, triggers a sizeable decline in GDP relative to the base case, which in turn affects the global GDP projections. However, in the longer term, the energy system is projected to recover from the decline in the fossil fuel energy industry, with low-carbon energy technologies being adopted at a larger scale contributing to a net increase in investments and a slightly higher GDP relative to the baseline (for instance, GDP for China in 2100 is expected to reach $25.8 trillion in the carbon price scenario compared with $25.2 trillion in the base case).[18] However, for other countries, such as the USA and India, the carbon price scenario results in GDP gains (albeit small), suggesting that the respective increase in real carbon prices leads to an overall increase in net electricity-sector investments, leading to higher exports and output.

4.2. Carbon-plus scenario

The implementation of the carbon-plus scenario entails five types of climate change mitigating measures (see Section 3.3), in addition to the rise in the global carbon price. In contrast to the carbon price scenario, there are real output gains compared with the base case. Global GDP gains stand at 1.1% relative to the baseline by 2050 (i.e. real global GDP rising to $104.1 trillion in 2050 in the carbon-plus scenario; see Figure 1 and Table 2). This is associated with achieving the LCS target of a 50% reduction in CO$_2$ emissions by 2050 relative to 2000 levels (see Table 2). The reductions in CO$_2$ emissions are distributed unevenly across countries, depending on their energy systems, opportunities for low-cost mitigation and the underemployed rural labour resources, with China cutting its emissions by 73% in 2050 whereas India continues to increase its emissions relative to 2000 by almost 60%, although the absolute increase is small (0.2 GtC). Compared with India, China uses much more coal and has greater opportunities to improve energy efficiency in many sectors. However, country results need to be assessed with caution, as there are large uncertainties surrounding the assumptions underpinning the respective regional outcomes (for example, the Indian economy is assumed to grow the most in the base case).

Higher economic growth due to mitigation is a key outcome of the E3MG model, as it argues that a decarbonization of the economy via induced technological change does not impede, but may in fact stimulate, long-term economic growth rates. In other words, as opposed to many findings in the literature,[19] and given underemployed resources in many countries, a less-carbonized economy may not cost more than a more-carbonized one. The overall GDP gains are explained in the E3MG model by the worldwide induced extra investments and/or R&D in alternative energy technologies as a consequence of both higher real carbon prices and the additional mitigation measures. Higher investments are associated with higher-quality and innovatory products (i.e. a switch to low-carbon technologies), leading to greater exports, which in turn are effectively matched by an increase in supply that is realized in the model through economies of scale, learning curves and higher productivity. The increase in global GDP relative to the carbon price scenario (by 3% in 2050) is distributed unevenly across countries and regions, with China and India benefiting the most, and the USA and Japan benefiting to a lesser extent (see Table 2). This reflects the significant economic potential that large, fast-growing, developing economies may have in the long run if change towards low-carbon technologies is consistently and effectively induced. For example, real GDP is expected to grow in the Cplus scenario in both China and India by more than a factor of ten in 2050 relative to 2000 levels (Table 2).

TABLE 2 Changes in CO_2 emissions and GDP in 2050

World regions – CO_2 emissions		EU25	USA	Japan	Russia	China	India	Rest of World	Global
Base 2050	Absolute level (GtC)	1.31	2.70	0.45	0.60	5.08	1.04	5.97	17.14
2050	% change	−39.2	−60.5	−32.4	−37.3	−82.7	−47.7	−49.1	−59.1
Cprice from base	Absolute change (GtC)	−0.51	−1.63	−0.14	−0.22	−4.20	−0.49	−2.93	−10.13
2050	% change	−57.5	−78.1	−51.5	−44.7	−92.6	−47.9	−73.9	−75.6
Cplus from base	Absolute change (GtC)	−0.75	−2.11	−0.23	−0.27	−4.70	−0.50	−4.41	−12.96
Cprice 2050	Absolute level (GtC)	0.79	1.07	0.30	0.38	0.88	0.54	3.04	7.00
2050	% change	−30.1	−44.6	−28.3	−11.8	−57.4	−0.5	−48.8	−40.3
Cplus from Cprice	Absolute change (GtC)	−0.24	−0.48	−0.09	−0.04	−0.50	−0.002	−1.48	−2.82
Year 2000	Absolute level (GtC)	1.21	1.72	0.37	0.48	1.39	0.34	2.81	8.32
2050	% change	−54.1	−65.7	−41.5	−31.2	−73.0	58.6	−44.5	−49.8
Cplus from 2000	Absolute change (GtC)	−0.65	−1.13	−0.15	−0.15	−1.01	0.2	−1.25	−4.14
Cplus 2050	Absolute level (GtC)	0.55	0.59	0.21	0.34	0.38	0.54	1.56	4.18
World regions – GDP, 2050		EU25	USA	Japan	Russia	China	India	Rest of World	Global
% change Cprice from base		−1.6	0.3	−1.7	4.4	−10.1	2.2	−0.7	−1.9
% change Cplus from base		2.0	1.0	−0.2	7.9	−3.4	10.1	1.6	1.1
% change Cplus from Cprice		3.6	0.8	1.4	3.3	7.5	7.7	2.4	3.0
% change Cplus from 2000		146.5	187.9	138.5	330.0	1014.7	1015.8	345.7	245.6

Source: E3MG2.3 SP7 baseline, LCS and LCP scenarios.

One major difference between China and India in their economic prospects over the next century, which helps to explain the differences in these results, is that China is expected to utilize all its underemployed rural resources by 2050, so that its growth will slow down. In contrast, it appears that the Indian economy has sufficient rural population and scope for rural–urban migration so that it can continue to grow rapidly through to 2100.

However, in order to disentangle the impacts stemming from mitigation in the carbon-plus scenario, we further identify, in a stepwise fashion in Table 3, the contribution of each measure introduced on a cumulative basis (in addition to the previous one) into the model. We find that, compared with the carbon price scenario, all the additional measures stimulate economic growth with the exception of the last measure, entailing bringing in earlier the carbon price of $100/tCO$_2$ (by 2030 instead of 2050). This is due to the fact that the additional measures all lead to new investment, simultaneously increasing demand and supply of products in the global economy. The pure carbon price measure increases the net reduction in overall investment, especially in China, and reduces the effects of the other additional measures. However, without the carbon price increase, there would be weak ITC and no extra revenues to recycle for the investment incentives, so the cost would have to come from general taxation or higher electricity prices, and the CO$_2$ target would become infeasible through rebound effects raising GDP and energy use. Greening et al. (2000) described the rebound effect in general, which occurs when improvements in energy efficiency lead to lower costs of energy use, and therefore higher use due to income and price effects, especially if energy is a large component of total costs. Evidence for strong rebound effects in developing countries is found in Roy (2000) for India, and in Glomsrod and Wei (2005) for China.

The additional reduction in carbon emissions by 2050 is small (around 2%) in the scenario entailing electricity technology subsidies, from a 16% fall in the Cprice scenario (equivalent to 7 GtC) to an 18% decrease in CO$_2$ emissions in the ADDelect case (equivalent to 6.8 GtC) – see Table 3. However, supporting the development of renewables and CCS technologies via subsidies fosters investment and growth in the world economy, the decrease in global GDP in the Cprice relative to the base case scenario being almost cancelled out (Table 2). The largest contribution, both in terms of CO$_2$ emission reductions and GDP gains, is attributed to the ADDtrans scenario, i.e. the early but gradual conversion of the transport vehicle fleet to an electrical plug-in system.[20] CO$_2$ emissions are reduced to around 40% in 2050 (relative to 2000 levels) and GDP gains are increased to 1.6% relative to the base case. This emphasizes the importance of addressing the transport sector through mitigation policies. The conversion to an electrical system will, in turn, increase the demand for electricity and accelerate both the rate of investment in low-carbon technologies, with positive impacts on exports and output, and the rate of emission reductions. The other two additional mitigation

TABLE 3 Decomposing the effects of the carbon-plus scenario

	Change in CO$_2$ emissions by 2050 relative to 2000 (%)	Changes in GDP relative to the base case, 2050 (%)
Base	+105.9	
Cprice	−15.9	−1.91
ADDelect	−18.2	−0.43
ADDtrans	−41.3	+1.60
ADDind	−43.3	+1.84
ADDdwell	−45.2	+1.92
Cplus	−49.8	+1.10

Source: E3MG2.3 SP7 baseline, LCS and LCP1 to eight scenarios.
Note: Cprice represents the carbon price scenario; the Cplus scenarios are explained in the text.

measures (ADDind and ADDdwell), targeting the energy-intensive industry and the household sector, also contribute to higher GDP growth and lower CO_2 emissions, albeit to a lesser extent. To achieve the 50% reduction in global CO_2 emissions in the full carbon-plus scenario, it was necessary to accelerate the rise in carbon price to $100/tCO_2$ by 2030, in addition to the sectoral mitigation measures. As this leads to an accelerated turnover of capital stock, it reduces the growth of GDP. However, the carbon-plus scenario consisting of a combination of the sectoral, technology-based mitigation measures and the accelerated rise in carbon price still leads to a 1.1% increase in global GDP relative to the base case.

This *Climate Policy* supplement includes a comparison of the E3MG approach and assumptions (and results) with that of other models (Strachan et al., 2008). Details of a similar comparison can be found in the IMCP study (Edenhofer et al., 2006).

5. Policy implications and conclusions

We argue that the scenarios described in this article suggest that achieving a global pathway to a low-carbon economy is feasible at modest carbon prices, but only by applying specific technology deployment policies in addition to the carbon price rising to $100/tCO_2$. This is line with the conclusions of the Stern Review (Stern, 2007), which argued that three complementary sets of policies are needed:

1. *Putting a price on carbon through taxes or trading schemes.* This is necessary but not sufficient, because of factors relating to:
   a. credibility of future carbon price
   b. uncertainties, risks of options and timescales
   c. underinvestment due to spillover effects.
2. *Accelerating technological innovation.* This needs to involve both:
   a. support for energy R&D (needs to double)
   b. creating markets and driving deployment (should increase between two and five times globally).
3. *Overcoming institutional and non-market barriers to adoption.*

Our results show that the largest contribution to reductions in global CO_2 emissions comes from the effect of the carbon price rising to $100/tCO_2$. The rising carbon price stimulates energy savings and the substitution of low-carbon products and technologies across all sectors. However, these carbon reductions are achieved at the cost of a 1.9% reduction in global GDP by 2050, because many potential energy savings, especially in buildings, are assumed to be unavailable through the usual market mechanisms. These are the so-called 'no-regrets' options with small or zero additional investment (e.g. low-energy light bulbs), compared with the alternative of high-energy-using buildings, requiring higher investment in power generation. To achieve the target of a 50% reduction in global CO_2 emissions, additional measures are needed to stimulate the innovation and deployment of low-carbon technologies, to reduce or remove barriers, and to bring forward the carbon price increases.

The most significant additional contribution to carbon reductions comes from measures to accelerate diffusion of electric vehicles through technology-based agreements to encourage decarbonizing of the transport sector. The widespread diffusion of electric vehicles is likely to require such technology-based agreements, alongside a rising carbon price, in order to stimulate the levels of investment needed in vehicles and infrastructure to overcome the lock-in of the current system (Foxon, 2003, 2007). A long-term framework comprising expectations of a rising carbon price and agreements to stimulate innovation and diffusion of low-carbon technologies could provide the level of certainty needed to stimulate such large-scale investments.

The additional technology-based measures in our model across the electricity, transport, industrial and housing sectors have a further significant benefit in our scenarios in that they lead to an overall rise in GDP relative to the base case. The combination of these measures with the rising carbon price achieves a 45% reduction in CO_2 emissions below 2000 levels. The additional acceleration of the rise in carbon price to $100/tCO_2$ by 2030 to achieve the 50% reduction in CO_2 emissions leads to a 1.1% increase in global GDP relative to the base case. The overall GDP gains are explained in the E3MG model by the worldwide induced extra investments and R&D in alternative energy technologies as a consequence of both higher real carbon prices and the additional mitigation measures. Higher investments are associated with higher quality and innovatory products (i.e. a switch to low-carbon technologies), leading to greater exports. These are effectively matched by an increase in supply that is realized through economies of scale, learning curves and higher productivity. In other words, we find that by including the effects of induced technological change, a transition to a low-carbon economy may not only be technologically feasible but may also be economically beneficial overall.

Notes

1. Since the choice of technologies is included within the model and affects energy demand and economic growth, then the model includes endogenous technological change (ETC). With ETC, further changes can generally be induced by economic policies, such as the $100/tCO_2$ carbon price and the policies implicit in the carbon-plus scenarios, so that these scenarios include (policy-) induced technological change (ITC).
2. The innovation externality arises when the social rate of return exceeds the private rate of return from R&D because innovators cannot capture all the benefits of their investment (see Jaffe et al., 2003).
3. See Clarke et al. (2006) for a review of the evidence.
4. '...the purpose of our simulation exercise was not to provide reliable predictions, but rather to identify the relevant processes and associated parameters of the system which need to be more closely investigated' (Weber et al., 2005, p.322).
5. The assumption of the representative agent, commonly adopted in equilibrium models, is that the behaviour of an economic group is adequately represented by that of a sample group, each of whose members have the identical characteristics of the average of the group. Barker and de-Ramon (2006) tested this assumption and showed that parameter estimates of the effects of technological change on industrial employment for 18 regions of Europe are significantly different across industries and regions. In other words, the assumption that each industry has the same responses as a hypothetical average (gathered from the literature) is unjustified and misleading.
6. The database was constructed by teams at Cambridge Econometrics (see www.camecon.com/).
7. This is in contrast to the mainstream equilibrium approach (see DeCanio, 2003, for a critique) adopted in most economic models of climate stabilization costs (see Weyant, 2004, for a discussion of technological change in this approach). Setterfield (1997) explicitly compares the approaches in modelling growth, and Barker (2004) compares them in modelling mitigation.
8. ADAM refers to the European Commission-funded project 'Adaptation and Mitigation Strategies supporting European Climate Policy' running from 2006 to 2009 that 4CMR–University of Cambridge is currently part of (see www.adamproject.eu/).
9. For a detailed description of the POLES (Prospective Outlook on Long-term Energy Systems) model, see http://webu2.upmf-grenoble.fr/iepe/Recherche/Recha5.html.
10. The main references to Pesaran's global model GVAR (developed in collaboration with the European Central Bank) are Dees et al. (2007) and Pesaran et al. (2004).
11. The GDP projections used in the EC's WETO-H2 (World Energy Technology Outlook 2050) by the POLES model rely on a neoclassical growth model, developed at CEPII (Centre d'Etudes Prospectives et d'Informations Internationales, Paris), with exogenous technological change and explicit consideration of human capital. Some of the long-term growth rates (particularly for India and China) have been slightly revised through expert consensus within the ADAM project framework.
12. For a detailed documentation of the TIMER (Targets Image Energy Regional) model, see www.mnp.nl/en/publications/2001/TheTargetsIMageEnergyRegionalTIMERModelTechnicalDocumentation.html.
13. See www.camecon.com/.

14. All values quoted are in year 2000 US$.
15. For example, if the model is run to project up to 2100, E3MG predicts that global real GDP in the carbon price scenario will increase at a faster rate compared with the base case by the end of the century (e.g. 200.1 (in the Cprice scenario) versus 194.7 trillion 2000 US$ in the base case by 2100).
16. CO_2 emissions refer to both emissions from energy and industry, and from deforestation. However, the latter are exogenous to E3MG and do not change across scenarios.
17. Although we have also included projections for Russia, these are to be treated with particular caution due to the poor quality of the data backing the modelling. The same caveat applies to China and other developing countries.
18. We assume that the capital costs associated with coal-fired power plants in China (and India) are half of the world levels due to the existence of a specialist market and large economies of scale in these countries. This implies that the conventional energy system in China (and India) faces lower investment costs than otherwise with the introduction of high carbon prices. In other words, it implies a smoother transition to low-carbon energy technologies. The GDP loss can be avoided if the investment resources could be planned in advance to be diverted from coal plant to additional renewable or other low-carbon plant. Nevertheless, a smoother transition in these countries would also require the building of a low-carbon technology capacity and an institutional framework to support this, which is in part dependent on international cooperation and effective technology transfer.
19. However, the IPCC AR4 (IPCC, 2007, SPM WG3 p.16) states that 'although most models show GDP losses, some show GDP gains because they assume that baselines are non-optimal and mitigation policies improve market efficiencies, or they assume that more technological change may be induced by mitigation policies. Examples of market inefficiencies include unemployed resources, distortionary taxes and/or subsidies.' At least 14 of the models considered in the AR4 have shown GDP above base for GHG mitigation at the global and national levels over different periods and under some combinations of assumptions.
20. Note that the model includes substantial energy efficiency improvements in transport in the base case due to technological change and regulation and additional improvements in the scenarios due to the effects of the carbon prices on fuel costs. The switch to electric vehicles through regulation and incentives is in addition to these effects.
21. All the detailed results in the tables should be taken as provisional and indicative only. Not all the estimated equations have been included in E3MG and a formal uncertainty analysis has yet to be performed. The projection for growth in global CO_2 energy-related emissions for 2030 is similar to that from the IEA (2007).

References

Anderson, D., Winne, S., 2004, *Modelling Innovation and Threshold Effects in Climate Change Mitigation*, Tyndall Working Paper 59 [available at www.tyndall.ac.uk/].

Awerbuch, S., 2006, 'Portfolio-based electricity generation planning: policy implications for renewables and energy security', *Mitigation and Adaptation Strategies for Global Change* 11(3), 693–710.

Barker, T., 2004, *Economic Theory and the Transition to Sustainability: A Comparison of General-Equilibrium and Space–Time–Economics Approaches*, Tyndall Working Paper 62, Tyndall Centre, University of East Anglia, UK, November.

Barker, T., de-Ramon, S.A., 2006, 'Testing the representative agent assumption: the distribution of parameters in a large-scale model of the EU 1972–1998', *Applied Economics Letters* 13(6), 395–398.

Barker, T., Haoran, P., Köhler, J., Warren, R., Winne, S., 2006, 'Decarbonising the global economy with induced technological change: scenarios to 2100 using E3MG', *Energy Journal* 27, 143–160.

Clarke, L., Weyant, J., 2002, 'Modeling-induced technological change: an overview', in: *Technological Change and the Environment*, Resources for the Future Press, Washington, DC, 320–363.

Clarke, L., Weyant, J., Birky, A., 2006, 'On the sources of technological change: assessing the evidence', *Energy Economics* 28, 579–595.

DeCanio, S., 2003, *Economic Models of Climate Change: A Critique*, Palgrave-Macmillan, New York.

Dees, S., di Mauro, F., Pesaran, M.H., Smith, L.V., 2007, 'Exploring the international linkages of the Euro area: a global VAR analysis', *Journal of Applied Econometrics* 22(1), 1–38.

Edenhofer, O., Lessman, K., Kemfert, C., Grubb, M., Köhler, J., 2006, 'Induced technological change: exploring its implications for the economics of atmospheric stabilisation. Synthesis Report from the Innovation Modeling Comparison Project', *Energy Journal* (Special Issue: *Endogenous Technological Change and the Economics of Atmospheric Stabilisation*), 1–51.

Fischer, C., 2003, 'Climate change policy choices and technical innovation', *Minerals and Energy* 18(2), 7–15.

Fischer, C., Newell, R., 2004, *Environmental and Technology Policies for Climate Change and Renewable Energy*, RFF Discussion Paper 04-05.

Foxon, T.J., 2003, *Inducing Innovation for a Low-carbon Future: Drivers, Barriers and Policies*, The Carbon Trust, London.

Foxon, T.J., 2007, 'Technological lock-in and the role of innovation', in: G. Atkinson, S. Dietz and E. Neumayer (eds), *Handbook of Sustainable Development*, Edward Elgar, Cheltenham, UK.

Glomsrod, S., Wei, T.Y., 2005, 'Coal cleaning: a viable strategy for reduced carbon emissions and improved environment in China?' *Energy Policy* 33, 525–542.

Greening, L., Greene, D.L, Difiglio, C., 2000, 'Energy efficiency and consumption: the rebound effect – a survey', *Energy Policy* 28, 389–401.

Grubb, M., Köhler, J., Anderson, D., 2002, 'Induced technical change in energy and environmental modelling: analytic approaches and policy implications', *Annual Review of Energy Environment* 27, 271–308.

IEA (International Energy Agency), 2007, *World Energy Outlook 2007: China and India Insights*, IEA, Paris.

IPCC, 2007, 'Summary for Policymakers', in: B. Metz, O.R. Davidson, P.R. Bosch, R. Dave, L.A. Meyer (eds), *Climate Change 2007: Mitigation*. Contribution of Working Group III to the Fourth Assessment Report of the Intergovernmental Panel on Climate Change, Cambridge University Press, Cambridge, UK.

Jaffe, A., Newell, R., Stavins, R., 2003, 'Technological change and the environment', in: K.-G. Mäler, J. Vincent (eds), *Handbook of Environmental Economics*, Elsevier, Amsterdam.

Jaffe, A., Newell, R., Stavins, R., 2005, 'A tale of two market failures: technology and environmental policy', *Ecological Economics* 54, 164–174.

Kaldor, N., 1957, 'A model of economic growth', *Economic Journal* 67(268), 591–624.

Kaldor, N., 1972, 'The irrelevance of equilibrium economics', *Economic Journal* 52, 1237–1255.

Kaldor, N., 1985, *Economics without Equilibrium*, University College Cardiff Press, Cardiff, UK.

McCombie, J.M., Thirlwall, A.P., 1994, *Economic Growth and the Balance of Payments Constraint*, St Martin's Press, New York.

McCombie, J.M., Thirlwall, A.P., 2004, *Essays on Balance of Payments Constrained Growth: Theory and Evidence*, Routledge Press, London and New York.

McDonald, A., Schrattenholzer, L., 2001, 'Learning rates for energy technologies', *Energy Policy* 29, 255–261.

McFarland, J., Reilly, J., Herzog, H., 2004, 'Representing energy technologies in top-down economic models using bottom-up information', *Energy Economics* 26(4), 685–707.

Masui, T., Hibino, G., Fujino, J., Matsuoka, Y., Kainuma, M., 2005, 'Carbon dioxide reduction potential and economic impacts in Japan: application of AIM', *Environmental Economics and Policy Studies* 7(3), 271–284.

Nakicenovic, N., Riahi, K., 2003, *Model Runs with MESSAGE in the Context of the Further Development of the Kyoto Protocol*, WGBU, Berlin.

Nordhaus, W.D., 2002, 'Modeling-induced innovation climate change policy', in: *Technological Change and the Environment*, Resources for the Future Press, Washington, DC, 182–209.

Pesaran, M.H., Schuermann, T., Weiner, S.M., 2004, 'Modelling regional interdependencies using a global error-correcting macroeconometric model', *Journal of Business and Economic Statistics* 22, 129–162

Roy, J., 2000, 'The rebound effect: some empirical evidence from India', *Energy Policy* 28, 433–438.

Scott, M., 1989, *A New View of Economic Growth*, Clarendon Press, Oxford, UK.

Setterfield, M., 1997, ''History versus equilibrium' and the theory of economic growth', *Cambridge Journal of Economics* 21, 365–78.

Setterfield, M., 2002, *The Economics of Demand-led Growth: Challenging the Supply-side Vision of the Long Run*, Edward Elgar, Cheltenham, UK.

Sijm, J.P.M., 2004, *Induced Technological Change and Spillovers in Climate Policy Modelling*, ECN, The Netherlands.

Stern, N., 2007, *The Economics of Climate Change: The Stern Review*, HM Treasury, London, and Cambridge University Press, Cambridge, UK.

Strachan, N., Foxon, T.J., Fujino, J., 2008, 'Policy implications from the Low-Carbon Society (LCS) modelling project', *Climate Policy* 8, Supplement, 2008, S17–S29.

Weber, M., Barth, V., Hasselmann, K., 2005, 'A multi-actor dynamic integrated assessment model (MADIAM) of induced technological change and sustainable economic growth', *Ecological Economics* 54(2–3), 306–327.

Weyant, J.P., 2004, 'Introduction and overview', *Energy Economics* (Special Issue: *EMF 19 Study Technology and Global Climate Change Policies*), 501–515.

Yang, Z., Nordhaus, W.D., 2006, 'Magnitude and direction of technological transfers for mitigating GHG emissions', *Energy Economics* 28(5–6), 730–741.

Young, A., 1928, 'Increasing returns and economic progress', *Economic Journal* 28 (152), 527–542.

■ research article

Global emission reductions through a sectoral intensity target scheme

KEIGO AKIMOTO*, FUMINORI SANO, JUNICHIRO ODA, TAKASHI HOMMA, ULLASH KUMAR ROUT, TOSHIMASA TOMODA

Research Institute of Innovative Technology for the Earth (RITE), 9-2 Kizugawadai, Kizugawa-shi, Kyoto 619-0292, Japan

If dangerous climatic change is to be avoided, all countries will need to contribute to reductions in greenhouse gas emissions on the basis of equity and in accordance with their common but differentiated responsibilities and respective capabilities. This article discusses the gap between the past (ideal) model analysis for emission reductions and realistic policies. A key requirement for successful policies is their acceptance by as many countries as possible and their ease of practical implementation. The sectoral intensity approach has been proposed for its focus on tangible, practical actions; however, its emission reduction effects have been said to be ambiguous and difficult to evaluate quantitatively. The effects of global emission reduction based upon a sectoral approach to energy and carbon intensity targets are evaluated using an energy systems model with a high regional resolution and a detailed description of technology. This analysis found that deep emission cuts can be achieved by a sectoral approach, provided that developed and developing countries collaborate towards emission cuts under the proposed framework. This framework has a higher potential for agreement by both developed and developing countries.

Keywords: emission reduction; intensity targets; low-carbon society; post-Kyoto; scenario modelling; sectoral approach; sectoral targets

Si le changement climatique dangereux peut être évité, tous les pays devront contribuer à la réduction des émissions de gaz à effet de serre sur la base du principe de l'équité et en accord avec leur responsabilités communes mais différenciées et leurs capacités respectives. Ce papier discute de l'écart entre l'analyse sur l'ancien modèle (idéal) de réduction des émissions et les politiques réalistes. Une exigence clé pour le succès des politiques est leur reconnaissance par le plus grand nombre de pays possible et leur facilité de mise en place en pratique. L'approche sectorielle d'intensité a été proposée pour son axe sur des actions tangibles et concrètes, cependant ses effets en terme de réductions en émissions ont été interprètés comme étant ambigus et difficiles à évaluer de manière quantitative. L'effet des réductions des émissions mondiales basée sur une approche sectorielle d'objectifs d'intensité en énergie ou en carbone est évalué à travers un modèle de systèmes énergétiques à haute résolution régionale et une description technologique détaillée. Cette analyse montre que d'importantes réductions d'émissions peuvent être réalisées par une approche sectorielle à condition que les pays développés et les pays en développement collaborent dans le but de réduire les émissions à l'intérieur du cadre proposé. Ce cadre présente un plus fort potentiel d'accord entre pays développés et pays en développement.

Mots clés: approche sectorielle; modélisation de scénarios; objectifs bases sur l'intensité; objectifs sectoriels; post-Kyoto; réduction des émissions; société sobre en carbone

■ *Corresponding author. E-mail*: aki@rite.or.jp

doi:10.3763/cpol.2007.0492 © 2008 Earthscan ISSN: 1469-3062 (print), 1752-7457 (online) www.climatepolicy.com

1. Introduction

The Fourth Assessment Report of the IPCC (AR4) states that 'most of the observed increase in globally averaged temperatures since the mid-20th century is very likely due to the observed increase in anthropogenic greenhouse gas concentrations' (IPCC WGI, 2007) and that 'a global assessment of data since 1970 has shown it is likely that anthropogenic warming has had a discernible influence on many physical and biological systems' (IPCC WGII, 2007). We must take strong action to reduce greenhouse gas emissions significantly.

In June 2007, the Group of Eight (G8) agreed to be 'committed to taking strong and early action to tackle climate change in order to stabilize greenhouse gas concentrations at a level that would prevent dangerous anthropogenic interference with the climate system', and to 'consider seriously the decisions (...) which include at least a halving of global emission by 2050' (G8 Summit, 2007). At present, there is no international agreement on the long-term stabilization level, and communication processes such as the US initiative as announced in May 2007 are desired to bring about a global agreement on the specific levels of the long-term targets.

The ministers of the member countries of the International Energy Agency (IEA) called 'on the IEA to promote the development of efficiency goals and action plans at all levels of government, making use of sector-specific benchmarking tools to bring energy efficiency to best-practice levels across the globe' and invited 'the IEA to evaluate and report on the energy efficiency progress in IEA Member and key non-Member countries' in May 2007 (IEA, 2007a).

In September 2007, the leaders of the Asia–Pacific Economic Cooperation (APEC) agreed 'to work to achieve a common understanding on a long-term inspirational global emissions reduction goal to pave the way for an effective post-2012 international arrangement' (APEC, 2007). In addition, they supported 'a flexible arrangement that recognizes diverse approaches, and supports practical actions and international cooperation across a broad range of areas relevant to climate change' and 'domestic actions which make measurable contributions to a shared global goal' (APEC, 2007). They also announced that the energy intensity in APEC regions would improve by at least 25% by 2030 with a peer-review system.

In response to these international mitigation actions on climate change, this article presents quantitative scenarios for ideal pathways by means of a simple model analysis and, in addition, realistic action plans towards deep emission reductions. The action plans have the aim of practical implementation possibilities by using an up-to-date global energy systems model with a high regional resolution and a detailed technology description. The model analysis in this article shows an assimilated scenario of a back-casting approach and realistic frameworks based on bottom-up actions through a sectoral approach.

2. Proposed post-Kyoto framework

A vision of long-term emission reductions is important for long-term investment and reforms in social systems. However, even if the emission reduction vision at global and/or national levels is shared, achieving the envisioned target is uncertain because appropriate actions are not easily planned for such emission reductions. In other words, visions should be discussed by a top-down or back-casting approach, but the real actions for emission reductions require a bottom-up approach.

In order to achieve deep emission cuts, this article proposes a sectoral approach for the world, which sets energy efficiency targets for appliances, cars, etc., and energy and/or CO_2 intensity targets for each industry sector. This framework is expected to facilitate the diffusion of high-energy-efficiency technologies and technological improvements. Tol (2002) proposed the 'technology protocol', which has three factors as parameters: a graduation income, a convergence rate, and an acceleration rate of technology improvements. Basically, the concept of the framework of this article is similar to Tol's in terms of technology orientation. However, the proposal in this study is more concrete and realistic; it is a sectoral and bottom-up approach, and it is based on the analysis of a highly technology-oriented model. The details of the proposed framework are as follows:

1. Energy and/or CO_2 intensity targets are set for sectors (e.g. electricity, iron and steel, cement, aluminium, paper and pulp, and chemical industry sectors).
2. Energy-efficiency targets are set for appliances and vehicles.
3. These intensity targets should be decided with the assistance of computer model analyses (as discussed below).
4. In order to promote technology diffusions in developing countries, a monetary fund should be established primarily by developed countries.
5. Although the proposed scheme of sectoral intensity improvements has a strong driving force to reduce CO_2 emissions, some additional policies are needed to encourage the energy-saving activities of residential and transport sectors.

The virtues of this framework are as follows:

■ It does not automatically require reduction in economic activities or shrinkage of the economic output, even when the projection of economic growth is incorrect and actual economic growth is higher than that of the projection. Furthermore, a cap on emissions would devastate the economy in the case of incorrect economic growth projections. On the other hand, if the actual economic growth is lower than the projection, the framework of sectoral intensity targets will not generate 'hot air', which typically slows down technology diffusions and improvements, while the cap scheme would.
■ The global cap-and-trade scheme must be strict and levy heavy penalties on participating entities in order to maintain a steady carbon market. The proposed framework can make the penalty instrument flexible, e.g. binding, pledge-and-review, and voluntary targets according to the regional and sectoral conditions.
■ It sets each intensity target based on the present levels of technology or energy efficiency, and therefore provides assurance that technological options exist for achieving the targets. In contrast, top-down or macro targets are usually economy-wide and determined without sectoral breakdown, and it is very difficult to allocate emissions to countries because of regional differences in economy, technological level, economic growth, etc.

These points are critical with respect to establishing a global emission reduction scheme, particularly for deep cuts. However, it is important to recognize that this framework has several shortcomings:

■ It would be difficult to set a proper intensity target by sector
■ The emission reductions are not achieved at the least cost
■ The amount of achievable emissions reduction is unknown without detailed analysis and evaluation, usually using a large-scale energy model.

However, the IEA (2007b) is investigating the current energy and/or CO_2 intensities by sector and is also developing indicators. In addition, the Asia–Pacific Partnership on Clean Development and Climate (AP6, 2007) is also working to investigate the differences in intensity and to establish common boundaries for intensity calculations across countries in the iron and steel and cement sectors, etc. Although den Elzen and Berk (2004) argued that 'while bottom-up approaches are concluded as being valuable components of a future climate regime [...] they do not seem to offer a real alternative to emission reduction and limitation targets, as they provide little certainty about the overall environmental effectiveness of climate policies', the work of the IEA and AP6 and the research described in this study will resolve the difficulty in setting the intensity targets and in estimating the resulting emission reduction amounts and their costs, thereby overcoming most of the shortcomings. Another shortcoming of the proposed framework is that larger actual emissions may occur, due to greater economic growth. However, the actual increase in emissions will be relatively small under such a target of deep emission reduction, which is designed to yield large improvements in intensity.

This article explores a sectoral intensity target by focusing on the quantitative evaluation of achievable emissions reductions, necessary technological means and reduction costs. It excludes the matters listed at (4) and (5) above: measures for technology transfer facilitation and measures for energy-saving promotion in residential, transport and other sectors.

3. The model

3.1. Framework of the model

This section briefly describes the model utilized in this study. The model, which we call DNE21+, is an energy systems model of intertemporal linear programming type. Its time span covers up to the middle of the 21st century, with representative time points at 2000, 2005, 2010, 2015, 2020, 2025, 2030, 2040 and 2050. In order to consider the existing and future regional differences, the model distinguishes 54 country-based regions. In addition, countries with large geographical areas, such as the USA, Canada, Australia, China, India, Brazil and Russia, are further disaggregated into 3–8 regions in order to take into account the transportation costs of energy and CO_2 in greater detail. The model encompasses 77 regions. The model attempts to determine the cost-effective measures under the given conditions, and their perfect foresight. To do this, the model minimizes the sum of discounted costs of world energy systems between 2000 and 2050 (a discount rate of 5% per year is adopted) for meeting various types of assumed production, services and energy demands. The model structure is similar to the IEA ETP model (IEA, 2006a); however, the divided regions in DNE21+ are greater in number than those in the ETP model and therefore are able to represent the regional differences more appropriately.

DNE21+ is a highly technology-oriented model, which considers not only energy supply technologies but also energy end-use technologies. An overview of the energy and CO_2 flows in DNE21+ is shown in Figure 1. Figure 2 shows a part of the assumed processes of high energy efficiency and a deep CO_2 emission cut in the iron and steel sector in DNE21+. Eight types of primary energy sources are explicitly modelled: natural gas, oil, coal, biomass,[1] hydro and geothermal, solar photovoltaics, wind, and nuclear. As technological options, various types of energy conversion technologies are explicitly modelled besides electricity generation. Some of them are oil refinery, natural gas liquefaction, natural gas reforming, coal gasification, water electrolysis, and methanol synthesis. The historical vintages of their technological facilities are taken into account. Carbon capture and storage (CCS) technologies are also modelled. Thanks to the modelling described above,

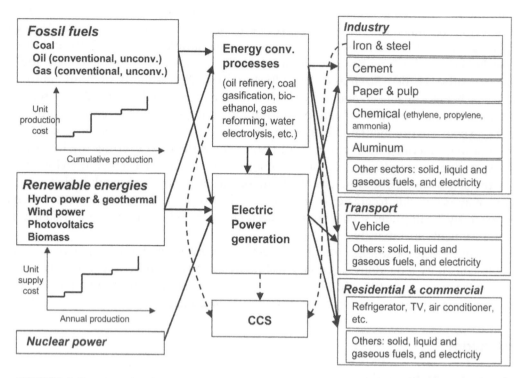

FIGURE 1 Overview of DNE21.+ model.

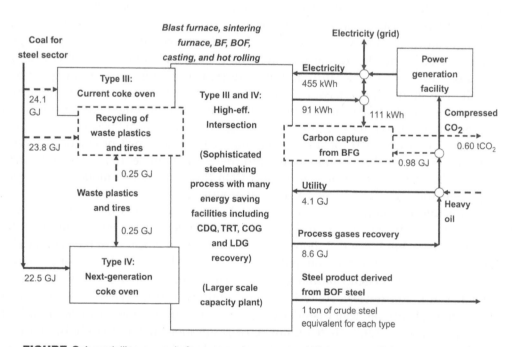

FIGURE 2 A modelling example for assumed processes of high energy efficiency and deep CO_2 emission cuts in the iron and steel industry (Oda et al., 2007).

DNE21+ is a strong and valuable support tool for the quantitative analyses and target settings for energy efficiency and/or carbon intensity by sector, appliance, etc., for the emission reduction framework proposed in Section 2.

The model utilized in this article treats technological change exogenously, and generally the costs of technologies are assumed to reduce over time and their performance to improve. The model evaluates only CO_2 emissions from fossil fuel combustion and does not cover the entire economy.

3.2. Key model assumptions

Future population and GDP are assumed, based on the IPCC SRES B2 (Nakicenovic et al., 2000). The global population and GDP growth between 2000 and 2050 are 0.86% and 2.4% per year, respectively. Most scenarios of future productions and activities, e.g. crude steel, cement, aluminium, paper and pulp production, passenger and freight transportation, are projected and used as assumptions based on the GDP per capita, the current trends, etc. Figure 3 shows the production of crude steel for selected model regions. Energy and CO_2 emission data in 2000 are calibrated using IEA statistics (IEA, 2006b, 2006c). These scenarios of future production and activities are kept fixed regardless of emission constraints.

Table 1 summarizes the world fossil fuel potential assumed in the model. The production costs of fossil fuel are assumed, based on the report of Rogner (1997). The FOB prices are calibrated in 2000 by using $57.5/toe, $199/toe ($29/bbl) and $110/toe for coal, crude oil and natural gas, respectively. Table 2 shows the assumed supply potentials and costs of hydro, wind, and photovoltaics in the world. The cost reductions of wind power and photovoltaics are assumed to be 1.0% and 3.4%, per year up to 2050, respectively (EPRI/DOE, 1997). The facility costs of fossil fuel power plants are calculated based on a report of NEA/IEA (1998), and the assumed efficiencies of coal, oil and natural gas fuel power plants are 22–55, 20–60 and 24–62% LHV, respectively; these include the regional differences and the technology improvements up to the year 2050, when their technology improvements are considered to be saturated. The model also assumes advanced technologies such as advanced coal fuel power plants (integrated coal gasification combined cycle (IGCC); integrated

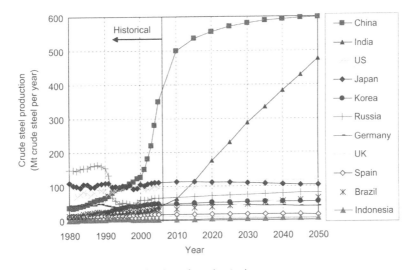

FIGURE 3 Assumed production of crude steel.

TABLE 1 Assumed fossil fuel potentials in the world (Gtoe; Gt of oil equivalent)

	Anthracite and bituminous		Sub-bituminous	Lignite
Coal	424		208	253
	Conventional			Unconventional
	Remaining Reserves	Undiscovered (Onshore)	Undiscovered (Offshore)	
Oil	137	60	44	2,342
Natural gas	132	59	52	19,594

Source: Rogner (1997), USGS (2000), WEC (2001).

TABLE 2 Assumed supply potentials and costs of hydro, wind and photovoltaics in the world

		Hydro power	Wind power	Photovoltaics
Supply potential (TWh/year)		14,400	12,000	1,271,000
	Year 2000	20–180	56–118	209–720
Supply costs ($/MWh)			1.0%/year reduction	3.4%/year reduction
	Year 2050	20–180	34–71	37–128

Note: The supply of electricity generated by wind power and photovoltaics to the electricity grid is limited to 15% of the total electricity supply for the stability of the grid. However, the use of a battery can extend this to an additional 15% of supply, and the electricity supply from photovoltaics directly to water electrolysis for hydrogen production is not limited. The price of the battery is assumed to be 375 and 7.6 $/MWh in 2000 and 2050, respectively.

coal gasification fuel cells (IGFC) etc.) with CO_2 capture, having the efficiency of 51% LHV; and advanced blast furnaces whose energy consumption is about 12.1 GJ for coal and 695 kWh for electricity per tonne of crude steel production. For further details, see Akimoto et al. (2004), Akimodo and Tomoda (2006) and Oda et al. (2007).

4. Model analysis

4.1. Simulation cases

Table 3 shows the simulation cases we studied for this research work.

Business-as-Usual (BaU) is the case where no specific policies for CO_2 emission reductions are adopted. However, currently, various types of policies for the reductions have already been implemented, particularly in developed countries, and thus the 'BaU' case is theoretical, and is used only for deciding the baseline of the mitigation costs for emission reduction within the model. This *Base* case considers major current climate policies and plans toward attaining the Kyoto target,[2] which includes global warming taxes in EU countries, nuclear power plans in Japan, etc. The *Carbon price* case assumes that the global carbon price is $10/tCO_2$ in 2013, which rises exponentially to $100/tCO_2$ in 2050. The *Vision 50/50* case assumes that global total CO_2 emissions in 2050 are halved relative to those in 2004, which corresponds to the target that was declared by

TABLE 3 Assumed simulation cases

Case	Assumed scenario
Business-as-Usual (BaU)	No specific policies for CO_2 emission reductions
Base	Current policies for CO_2 emission reductions are taken into account
Carbon price	10\$/tCO_2 in 2013 rising exponentially to 100\$/tCO_2 in 2050
Vision 50/50	Half of CO_2 emission reductions by 2050 relative to those in 2004: the year of the global emission peak is around 2020
Sectoral approach toward 50/50	Energy and carbon intensity targets by sector, appliances, and vehicles, which correspond to the 'Vision 50/50' case

Note: Policies for CO_2 emission reductions that are currently adopted or are to be adopted in the near future are taken into account for the three policy cases: 'Carbon price', 'Vision 50/50' and 'Sectoral approach toward 50/50'.

the last G8 summit to 'consider seriously'. The latter two scenarios assume ideal situations in which the marginal reduction costs are uniform among the regions and sectors throughout the world. However, this assumption does not reflect the real world; any cap-and-trade scheme will not realize uniform marginal costs in practice. The reason is that specific policies and measures must be planned and implemented in different sectors of different regions, and the marginal costs will not become common across sectors, even under a global cap-and-trade scheme. The only possibility is a uniform international carbon tax scheme; however, this is unrealistic in many respects. Most developing countries will never participate in such a scheme. Hence, this article conducted an additional model run, the *Sectoral approach toward 50/50* case, which assumes a much more realistic scheme; this is proposed in Section 2. The specific intensity targets by sector were determined so that a comparable amount of emission reductions may be achieved as for the 'Vision 50/50' case and at a reasonably low cost.

4.2. Model results and discussions

4.2.1. Overview of the simulation results

Global net CO_2 emissions from fossil fuel combustion for the assumed cases are shown in Table 4. While the CO_2 emission in the 'BaU' case will reach 60 Gt in 2050, CO_2 emission in the 'Carbon

TABLE 4 Global net CO_2 emissions from fossil fuel combustion (GtCO_2)

	Y2000	Y2010	Y2020	Y2030	Y2040	Y2050
Business-as-Usual (BaU)	22.7	31.0	38.4	47.1	54.4	60.0
Base	–	29.3	35.9	44.1	51.5	58.0
Carbon price	–	28.8	29.2	30.1	27.9	21.7
Vision 50/50	–	29.0	30.7	29.0	23.7	13.0
Sectoral approach toward 50/50	–	29.0	31.9	32.4	30.1	22.5

price' and 'Sectoral approach toward 50/50' cases will be 21.7 and 22.5 Gt, respectively, in 2050. The emission in the 'Sectoral approach toward 50/50' case is larger than in the 'Vision 50/50' case, but is approximately the same as in the 'Carbon price' case. The global emission reaches its peak in around 2030 and thereafter decreases. The proposed sectoral approach expects a large emission reduction (approximately 15 and 38 Gt in 2030 and 2050, respectively) indicating specific policies and technology measures required to achieve it.

The marginal cost of carbon emission reduction in 2050 in the 'Vision 50/50' case is $205/tCO_2$ for the world. This cost is an ideal one, on the assumption of minimum global cost, which requires that the marginal cost is uniform across countries and sectors; however, it is particularly high due to the deep emission reduction. The common marginal cost cannot be obtained for the 'Base' case and 'Sectoral approach toward 50/50' case because the cost is different across sectors and regions.

The averaged costs of CO_2 emission reduction are shown in Table 5. 'Carbon price' and 'Vision 50/50' cases assume an ideal world, and the costs are the least for the same amount of the emission reduction; however, these schemes are quite unrealistic in terms of, for instance, 'differentiated responsibilities' and 'respective capabilities'. The global CO_2 emission reduction in the 'Sectoral approach toward 50/50' case is slightly smaller than that in the 'Carbon price' and 'Vision 50/50' cases. Note that the average cost in the 'Sectoral approach toward 50/50' is almost the same as in the 'Carbon price' and 'Vision 50/50' by 2030. This is because its intensity targets are decided considering the model analysis results of 'Vision 50/50' where minimum costs are ensured. Note that as the facilities' vintages are taken into account in the model, the model analysis of cost minimization provides the solution based on the current regional differences in technology level. The intensity targets cannot be uniquely determined for a given amount of emission reduction and the adopted ones should be regarded as an example with emphasis on the cost-effectiveness. The proposed and analysed sectoral approaches described in this article can be an effective strategy towards achieving a low-carbon society. However, the difference in the emission reduction between the 'Vision 50/50' and 'Sectoral approach toward 50/50' cases is relatively large in 2050. This is discussed in greater detail in subsequent sections.

TABLE 5 Averaged costs of CO_2 emission reduction from the 'BaU' case ($/tCO_2$)

		Y2020	Y2030	Y2050
Carbon price	World	17.6	19.2	31.8
	Annex I	29.3	21.9	22.2
	Non-Annex I	7.7	11.4	30.7
Vision 50/50	World	18.9	21.2	49.6
	Annex I	30.7	22.9	33.5
	Non-Annex I	7.3	14.4	52.6
Sectoral approach toward 50/50	World	21.7	20.7	45.4
	Annex I	34.9	25.0	32.8
	Non-Annex I	7.1	11.4	45.3

TABLE 6 Overview of the intensity target of the 'Sectoral approach toward 50/50' case

		Y2020	Y2030	Y2050
Power sector	Annex I	1.12	0.73	−0.37
	Non-Annex I	1.13	0.62	−0.45
Iron and steel	Annex I	0.92	0.90	0.65
	Non-Annex I	1.05	0.94	0.73
Cement	Annex I	1.05	1.03	0.89
	Non-Annex I	1.33	1.23	0.93
Small passenger car	Annex I	0.73	0.57	0.39
	Non-Annex I	1.11	0.88	0.37
Bus	Annex I	0.79	0.57	0.32
	Non-Annex I	0.89	0.67	0.33

Note: The numbers designate the intensities relative to the current levels in Japan. The intensity is with respect to CO_2 emission and energy consumption for power and other energy conversion sectors and for all the other sectors, respectively (e.g. gCO_2 per kWh of electricity output in the power sector; toe per tonne of crude steel production in the iron and steel sector; toe per tonne of clinker production in the cement sector).

4.2.2. Overview of the intensity target of the 'Sectoral approach toward 50/50' case

An overview of the intensity target of the 'Sectoral approach toward 50/50' case is illustrated in Table 6, which shows the intensity targets aggregated for two regions and for the selected sectors; however, in fact, the intensity targets are established for power sectors, other energy conversions, iron and steel, cement, aluminium, paper and pulp, and chemical industries, vehicles (car, bus and truck), and appliances (TV, air conditioner and refrigerator) by region and by time for all the 54 regions. The carbon intensity in the power sector has to be negative in 2050 from the viewpoint of cost-effectiveness, as suggested by the analysis result for 'Vision 50/50', where all the fossil fuels and biomass power plants have CO_2 capture facilities.

4.2.3. Technological measures

The primary energy supply for the assumed cases is shown in Figure 4. Fossil fuels dominate the primary energy in the 'BaU' and 'Base' cases throughout the evaluation periods. Even in the 'Carbon price', 'Vision 50/50' and 'Sectoral approach toward 50/50' cases, fossil fuels remain dominant in the primary energy; however, the share of fossil fuels decreases and that of nuclear power, wind power, photovoltaics and bioenergy increases considerably. The use of a large amount of CCS is cost-effective in these cases. The amount of CO_2 storage in 2050 is 2.5, 15.2, 18.7 and 17.3 $GtCO_2$ in the 'BaU', 'Carbon price', 'Vision 50/50' and 'Sectoral approach toward 50/50' cases, respectively.

Emission reduction amounts by sector and technological option for the 'Sectoral approach toward 50/50' are shown in Table 7. The energy efficiencies are improved, and the CO_2 storage by enhanced oil recovery operations is cost-effective even in the 'BaU' case, as might be expected. Note that the reductions in Table 7 are shown relative to the emission in the 'BaU' case. Large emission reductions due to energy saving in many sectors, nuclear power introduction, and fuel

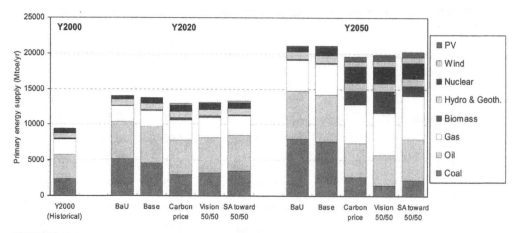

FIGURE 4 Global primary energy supply in 2000, 2020 and 2050.

TABLE 7 CO$_2$ emission reduction amounts by sector and technological option for 'Sectoral approach toward 50/50' case (GtCO$_2$)

		Y2020	Y2030	Y2050
Electric power	Energy saving	1.87	3.48	3.06
	Fuel switching among fossil fuels	1.26	2.52	1.69
	Nuclear power	1.63	3.79	5.54
	Hydro and geothermal	0.16	0.17	0.10
	Biomass	0.04	0.08	5.00
	Wind power	0.15	0.39	0.78
	Photovoltaics	0.00	0.00	0.96
	CCS	0.27	2.09	10.38
Other energy conversion		−0.02	−0.18	0.69
Iron and steel	Energy saving and fuel switching	0.05	0.08	2.66
	CCS	0.00	0.00	0.66
Cement		0.01	0.05	0.32
Paper and pulp		0.00	0.00	0.00
Chemical (ethylene, propylene and ammonia)		0.12	0.30	0.69
Aluminium		0.01	0.03	0.07
Other industrial sectors		0.38	0.64	1.23
Transportation		0.47	0.92	3.17
Residential and commercial		0.13	0.29	0.64

Note: The emission reduction effects of CCS are estimated as the net effects, taking into account the emission increase effect due to increased energy consumption in CO$_2$ capture and other processes.

switching from coal to natural gas in the power sector are expected in 2020; CCS is also expected to contribute greatly to emission reductions after 2030; renewable uses in power sectors and innovative technologies in many sectors, e.g. advanced nuclear power, direct iron reduction utilizing hydrogen which is generated from fossil fuels by using CCS, plug-in hybrid and fuel-cell cars, are also required to contribute to deep emission reductions by 2050.

The reductions, particularly in other industrial sectors for which the intensity targets were not applied, transportation sectors, and residential and commercial sectors in the 'Sectoral approach toward 50/50' case are smaller than in the 'Vision 50/50' case. Additional policy measures, e.g. for social system changes in transportation and residential and commercial sectors, will be required for further emission reductions. The reason for setting energy-intensity targets for sectors except for energy conversion sectors is the consideration of acceptability by the whole world, especially by developing countries. If carbon-intensity targets were assumed for all the sectors, larger emission reductions would be brought about in 2050 at the expense of lower acceptability of the scheme by developing countries.

4.2.4. Discussions on policy implications

The above simulation results show that the proposed sectoral framework, where energy-efficiency targets are set for appliances, cars, etc., and energy and/or CO_2 intensity targets for each industry sector, can achieve deep emission cuts, indicating the necessary technological options. The framework provides technology perspectives towards emission reductions during the next 10–20 years at least, and is also able to lead to pathways towards achieving a low-carbon society by the middle of this century. The achievement of the deep emission cuts will be technologically feasible and will be achieved even under the realistic framework of sectoral intensity targets that will be accepted with less difficulty by both developed and developing countries. However, we assumed in the DN21+ model considerable technology improvements in photovoltaics, wind power, batteries, CO_2 capture, nuclear power, bioenergy technologies, and hydrogen-related technologies, etc., which might be regarded as a little too optimistic for some technology areas. Despite the possibly optimistic assumptions on technological progress in the future; as cost reductions are their major feature, the achievement of the deep emission cuts requires particularly high emission reduction costs for the world. For this reason, strong and early R&D actions to facilitate the cost reductions in CO_2 emission reduction technologies are also desired. In addition, conditions for wide acceptance of nuclear power, CCS, and hydrogen technologies will have to be generated. The achievement of a low-carbon society is a difficult and narrow pathway. However, assimilation of the vision towards a low-carbon society and a framework to induce the bottom-up actions plus enthusiastic and committed international R&D efforts would overcome the difficulty and make the pathway more solid.

5. Conclusions

Both a vision of a low-carbon society and proper frameworks are required to address significant CO_2 reductions. The proposed sectoral approach of energy and carbon intensities has a high potential to realize deep emission reductions and to obtain the agreement and participation of both developed and developing countries. One of the problems with this framework is the unpredictability of the achievable emission reductions without laborious analyses using large-scale energy models. The other is the cost-effectiveness. We developed a technology-oriented global energy model with a high regional resolution, DNE21+, mainly for addressing these problems.

A quantitative evaluation of achievable emission reductions was made under a scheme of sectoral intensity targets using DNE21+. The use of such models also ensures the cost-effectiveness of this framework; we can set intensity targets, considering the model analysis results of cost-minimization for a certain amount of emission reductions. On the other hand, the global cap-and-trade scheme might anticipate the firm emission reduction to be achieved when the penalty for failing to meet the targets is heavy enough. However, realization of the cap-and-trade scheme requires specific sectoral actions at national level, which is usually outside the discussion in the negotiation processes of the cap-and-trade, and institutions must be planned and implemented in each country in accordance with the cap. Thus, whether or not the nationally allocated emission reductions are really achieved will not necessarily be certain. (Note the recent declaration of the Canadian government that Canada will fail to meet its agreed emission target set by the Kyoto Protocol.) In contrast, the framework proposed in this article promotes emission reductions, indicating proper actions at implementation levels for all the regions.

The roadway towards a low-carbon society is long and difficult. Despite the expected large emission reductions by the proposed framework, considerable R&D efforts are also required for the achievement of a low-carbon society. In addition, the burdens and the costs of the emission reductions across countries need to be discussed – although this article did not focus on this important issue. We have just shown that a sectoral intensity target scheme could bring about a large amount of emission reduction at reasonably low costs when the scheme is well designed. The equity issue will make it even more difficult to design internationally agreed frameworks, particularly for deep emission reductions, which reasonably require burden sharing by developing countries. One of the possible ways to alleviate this equity problem may be setting energy-intensity targets for developing countries and carbon-intensity targets for developed countries during the next decade or few decades. The proposed framework, as it alleviates the impacts of emission reductions on economic growth and interferes less with the international competitiveness of sectors, will attain this international agreement with less difficulty.

More research is needed in the near future regarding the sectoral intensity framework with various levels of intensity targets, taking into consideration 'differentiated responsibilities' and 'respective capabilities'.

Acknowledgements

The authors would like to thank Professor Yoichi Kaya (RITE) and many other advisers. This study is in part supported by the New Energy and Industry Technology Development Organization (NEDO), Japan.

Notes

1. Only commercial use is treated in DNE21+.
2. IEA World Energy Outlook (WEO) considers government policies and measures that were enacted or adopted by mid-2006 for its 'Reference scenario' (IEA, 2006d). The 'Base' case in DNE21+ has a similar treatment.

References

Akimoto, K., Tomoda, T., 2006, *Costs and Technology Role for Different Levels of CO$_2$ Concentration Stabilization: Avoiding Dangerous Climate Change*, Cambridge University Press, Cambridge, UK.

Akimoto, K., Homma, T., Kosugi, T., Li, X., Tomoda, T., Fujii, Y., 2004, 'Role of CO_2 sequestration by country for global warming mitigation after 2013', in: *Proceedings of the 7th International Conference on Greenhouse Gas Control Technologies, Vol.1: Peer-Reviewed Papers and Plenary Presentations*.

AP6, 2007, *The Asia–Pacific Partnership on Clean Development and Climate* [available at www.asiapacificpartnership.org/].

APEC, 2007, *Sydney APEC Leaders' Declaration on Climate Change, Energy Security and Clean Development* [available at www.apec.org/apec/leaders__declarations/2007/aelm_climatechange.html].

den Elzen, M.G.J., Berk, M.M., 2004, *Bottom-up Approaches for Defining Future Climate Mitigation Commitments*, RIVM Report 728001029.

EPRI/DOE, 1997, *Renewable Energy Technology Characterizations*, EPRI Topical Report TR-109496, EPRI, CA.

G8 Summit, 2007, *Growth and Responsibility in the World Economy* [available at www.g-8.de/Webs/G8/EN/G8Summit/SummitDocuments/summit-documents.html].

IEA, 2006a, *Energy Technology Perspectives*, OECD/IEA.

IEA, 2006b, *Energy Balances of OECD/Non-OECD*, OECD/IEA.

IEA, 2006c, *CO_2 Emissions from Fuel Combustion*, OECD/IEA.

IEA, 2006d, *World Energy Outlook*, OECD/IEA.

IEA, 2007a, *Communique: Meeting of the Governing Board at Ministerial Level* [available at www.iea.org/Textbase/press/pressdetail.asp?PRESS_REL_ID=225].

IEA, 2007b, *Tracking Industrial Energy Efficiency and CO_2 Emissions*, OECD/IEA.

IPCC WGI, 2007, *Climate Change 2007: The Physical Science Basis*. Contribution of Working Group I to the Fourth Assessment Report of the IPCC, Cambridge University Press, Cambridge, UK.

IPCC WGII, 2007, *Climate Change 2007: Impacts, Adaptation and Vulnerability*. Contribution of Working Group II to the Fourth Assessment Report of the IPCC, Cambridge University Press, Cambridge, UK.

Nakicenovic, N., et al. (eds), 2000, *Special Report on Emissions Scenarios*, Cambridge University Press, Cambridge, UK.

NEA/IEA, 1998, *Projected Costs of Generating Electricity: Update 1998*, OECD, Paris.

Oda, J., Akimoto, K., Sano, F., Tomoda, T., 2007, 'Diffusion of energy efficient technologies and CO_2 emission reductions in iron and steel sector', *Energy Economics* 29, 868–888.

Rogner, H.-H., 1997, 'An assessment of world hydrocarbon resources', *Annual Review of Energy and Environment* 22, 217–262.

Tol, R.S.J., 2002, *Technology Protocols for Climate Change: An Application of FUND* [available at www.uni-hamburg.de/Wiss/FB/15/Sustainability/bat.pdf].

USGS, 2000, *U.S. Geological Survey World Petroleum Assessment 2000: Description and Results* [available at http://greenwood.cr.usgs.gov/energy/WorldEnergy/DDS-60/].

WEC, 2001, *Survey of Energy Resources 2001* (CD-ROM), World Energy Council, London.

climate policy

■ esearch article

A global perspective to achieve a low-carbon society (LCS): scenario analysis with the ETSAP-TIAM model

UWE REMME*, MARKUS BLESL

Institute of Energy Economics and the Rational Use of Energy (IER), University of Stuttgart, Hessbruehlstrasse 49a, 70565 Stuttgart, Germany

Global warming caused by an increase of the concentration of greenhouse gases (GHG) from human activities is threatening the natural and human environment by extinction of species, sea-level rise, and change in the availability of water or increased frequency of extreme weather events. Within the UK–Japan Low Carbon Society (LCS) project, the global, technology-rich ETSAP-TIAM model has been applied to analyse, by means of a scenario analysis, strategies to realize deep GHG emission reductions on a global level. The scenario analysis shows that, without any explicit abatement efforts, energy-related carbon dioxide (CO_2) emissions are estimated to double by the middle of this century compared with the year 2000. With CO_2 abatement measures being equivalent to a CO_2 price of up to \$100/t in 2050, CO_2 emissions can be reduced by 23% relative to levels in 2000. Further efforts to halve CO_2 emissions in 2050 relative to 2000 levels can be achieved in a future energy system characterized (besides efficiency improvements and increased use of renewables, especially biomass) by an almost entirely decarbonized power generation sector (through carbon capture and storage power plants, renewable technologies and nuclear power), which provides electricity as the major final energy carrier to the end-use sectors. Since the majority of the emission reductions occur in the present developing countries, cooperation between developed and developing countries in the implementation of these measures is indispensable in order to realize these ambitious reduction targets.

Keywords: carbon pricing; climate change; emission reduction; energy policy; energy technological models; energy systems; global energy model; low-carbon society

*Le réchauffement planetaire cause par une augmentation de la concentration de gaz à effet de serre (GES) issus des activités humaines menace l'environnement naturel et humain par des phenomenes d''extinction d'espèces, une hausse du niveau de la mer, des changements dans l'approvisionnement en eau, et dans la fréquence accrue d'evenements climatiques extrêmes. Dans le cadre du projet Japon-Royaume-Uni de société sobre en carbone (LCS), le modèle mondial riche en technologie ETSAP-TIAM a été applique pour analyser des stratégies de réductions d'émissions de GES profondes à l'échelle mondiale. L'analyse des scenarios montre qu'a defaut d'efforts de réductions formules, les émissions de CO_2 issues de l'énergie doubleraient au milieu de ce siècle compare a leur niveau de 2000. En appliquant des mesures de réduction du CO_2 équivalent a un prix du CO_2 allant jusqu'à 100 \$/t en 2050, les émissions de CO_2 pourraient être réduites de 23% compare a leur niveau de 2000. Des efforts plus soutenus de réduire de moitie les émissions de CO_2 de 2050 relatif a leur niveau de 2000 pourraient être obtenus dans le cadre d'un système énergétique futur caractérisé (ainsi que par le biais de gains d'efficacité et une utilisation accrue de l'énergie renouvelable, surtout la biomasse), par une production d'énergie presqu'entièrement decarbonée (centrales de capture et stockage du carbone, technologies renouvelables et énergie nucléaire), fournissant de l'électricité en tant que produit final pour l'utilisateur. Vu que la majorité des réductions d'émissions se déroule dans les pays en développement actuels, une coopération entre pays développés et pays en développement dans la mise en place de ces mesures est indispensable dans le but de réaliser ces objectifs de réduction ambitieux.

Mots clés: changement du climat; fixation du prix du carbone; modélisation de technologies d'énergie; modélisation énergétique mondiale; politique énergétique; réduction des emissions; société sobre en carbone; systèmes énergétiques

■ *Corresponding author. E-mail:* uwe.remme@ier.uni-stuttgart.de

doi:10.3763/cpol.2007.0493 © 2008 Earthscan ISSN: 1469-3062 (print), 1752-7457 (online) www.climatepolicy.com

1. Introduction

Combating climate change and mitigating the emission of greenhouse gases (GHG) is a global challenge. In the past, developed countries have been major global emitters. In the future, economic growth in developing countries and regions, such as in China and India, is expected to continue the trend of increasing emissions of the major GHG, carbon dioxide (CO_2) emissions in these regions, if no actions are taken. Between 1990 and 2004, total energy-related CO_2 emissions in non-OECD countries have grown by 41%, compared with 17% in OECD countries (IEA, 2006). Projections estimate that developing countries in Africa, Asia and Latin America, as well as Small Island States, will be hit earliest and hardest by climate change (Yohe et al., 2007). Against this background, various proposals for reducing GHG beyond Kyoto are under discussion. At the G8 Summit in Germany in June 2007, all parties agreed to 'seriously consider' cutting global emissions by at least 50% relative to 2000 levels by 2050. Earlier that year the EU proposed to limit global warming to 2°C relative to pre-industrial levels and agreed to reduce CO_2 emissions by 20% by 2020 relative to 1990, even in the absence of international agreements (EC, 2007). Individual countries have already announced ambitious reductions goals: the UK government has already proposed a 60% cut in CO_2 emissions by 2050, and the German government announced in August 2007 a possible package of measures to reduce emissions by 40% by 2020.

Within the LCS scenario analysis, the global ETSAP-TIAM model[1] (TIMES Integrated Analysis Model) is used to analyse the achievement of deep emission reductions from a global perspective. A particular strength of this modelling framework is the technology-rich depiction of the energy system, taking into account the interconnections between the different sectors of the system. This allows a coherent identification of cost-effective mitigation strategies in the 15 different world regions depicted in the model. In this article, a short overview of the TIAM model is given, which is followed by a discussion of the scenarios analysed as well as the technology conclusions derived. The LCS scenario analysis is based on three different types of scenarios: a *Base* development without any explicit CO_2 mitigation efforts (BASE), a *CO_2 price* scenario with a CO_2 incentive reaching \$100/tCO_2 in 2050 (C10), and a *CarbonPlus* scenario enforcing a 50% reduction of CO_2 emission by 2050 (CPLUS).

2. The global ETSAP-TIAM model

The energy system model TIAM (Loulou and Labriet, 2007a, 2007b), which is based on the model generator TIMES (Loulou et al., 2005), is a technology-rich, bottom-up model depicting the global energy system. The model covers the time horizon from 2000 to 2100. The analysis presented in this article focuses, however, on the time span from 2000 to 2050. In the TIAM model, the world is divided into 15 world regions, which are the USA, Canada, Mexico, Latin America, Western Europe, Eastern Europe, the Former Soviet Union, Africa, the Middle East, India, China, Japan, South Korea, Other Developing Asia, and Australia/New Zealand. The primary energy resources and the petroleum processing sector are further divided into OPEC and non-OPEC sub-regions; while in the residential and commercial sector up to four different sub-regional areas for capturing different heating and cooling demands are distinguished. The world regions are linked through the trade in crude oil, hard coal, pipeline gas, LNG (liquefied natural gas), petroleum products (diesel, gasoline, naphtha, heavy fuel oil) and emission permits. Therefore, the physical trade activities are described by pipelines or tankers, taking into account the existing capacities and their technical and economic characteristics as well as new trade options and their investment costs. On the resource side, the conventional and unconventional oil and gas reserves and resources in the different regions, as well as various enhanced recovery methods, are included in the model (oil: extra-heavy oil, oil shale, tar sands; natural gas: coal-bed methane, aquifer gas, tight gas). Coal accumulations of hard coal and lignite are distinguished in reserves and resources. In addition, renewable energy sources and their potentials, as well as alternative technologies for synthetic fuels (e.g. coal-to-liquid, gas-to-liquid) and different pathways for hydrogen production, are considered in the supply side of the model. In each region, the TIAM model describes the entire energy system with regard to all essential current and future energy technologies from the primary energy supply through the processing, conversion, transport, distribution of energy carriers to the end-use sectors and the useful energy demand (Figure 1).

The model horizon considered here from 2000 until 2050 is divided into periods of 5 years duration, being represented by an average year. To reflect seasonal or diurnal variations in demand or supply, the representative year in TIAM is divided into three seasons and two daily time segments.

An integrated, simplified climate module in TIMES allows the determination of the CO_2 concentration changes in the atmosphere and thus the estimation of the induced global temperature changes. In addition to CO_2, TIAM also balances the greenhouse gases N_2O and CH_4. Marginal abatement curves for the process-related emissions of the latter two are implemented, whereas for CO_2 capturing at power, synthetic fuel and hydrogen production plants followed by storage in geological formations is being considered. A stochastic version of the TIAM model provides the option to analyse hedging strategies regarding uncertainties in future emission targets and demands. Useful energy demands can be formulated in a price-elastic manner or can be determined in a simple macroeconomic extension of the TIMES model, which describes the remaining part of the

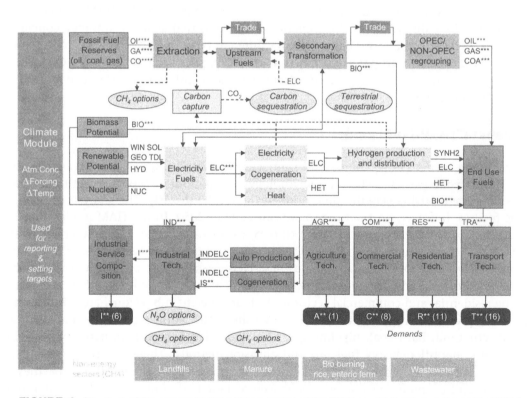

FIGURE 1 Structure of the energy sector within a region of the TIAM model (Loulou and Labriet, 2007a).

economy outside of the energy sector by a single-sector computable general equilibrium (CGE) model. The latter two options, however, have not been considered in this scenario analysis, which means that the assumed economic development, and thus the useful energy demand, is the same across all scenarios. The implication for the scenario analysis is that CO_2 reductions realized by reduced economic development or behavioural changes are excluded; the reduction targets have to be achieved within the energy sector. Nevertheless, useful energy demand can be reduced by technological conservation measures, such as reduced heating demand by better insulation of the building shell (Ürge-Vorsatz et al., 2007a, 2007b).

The TIAM model, which in its standard formulation is a linear programming model, minimizes the total discounted energy system costs over the entire model horizon. Results of the optimization are the structure of the energy system, i.e. type and capacity of the energy technologies, the energy consumption by fuel, the development of emissions, the energy trade flows between world regions, as well as the necessary transport capacities, the energy system costs, the long-term prices for the energy carriers, as well as the marginal costs of environmental measures, e.g. marginal abatement costs for GHG reduction targets.

Input data and assumptions of the TIAM model can be broadly categorized into parameters defining the useful energy demand, describing the available fossil resources as well as renewable potentials and data characterizing the energy technologies in terms of costs and their technical characteristics. The energy service demands are projected from historic figures in demand functions, using projection for drivers (e.g. population or gross domestic product) and elasticities to reflect the decoupling of demands and their respective drivers. Documentation on the demands and the other input data assumptions of the TIAM model can be found at ETSAP (2007) and in Remme et al. (2007).

One of the strengths of the TIAM model lies in the capability to analyse climate mitigation strategies in the world energy system in a consistent way based on a detailed technology description of the energy sector and its internal interdependencies. However, due to the partial equilibrium nature of the model, interactions with the remaining economy can only be considered in form of elastic demand curves or by the linkage with a CGE model as mentioned above. Despite the global nature of TIAM, the division into major countries and world regions, respectively, still offers the possibility of capturing regional differences in terms of currently existing energy infrastructure, renewable potentials or future economic development. In most cases, individual countries are not described explicitly in the model,[2] in order to keep the model computationally manageable. A consequence is that individual national conditions (e.g. on the demand side as detailed building stock) and policies and their impacts cannot be described in detail in TIAM, but are only captured in aggregate form within the larger region. However, one can envision a sequence of model runs and information exchanges between the global TIAM model and national models: with the global model providing framework assumptions as global energy prices for the national models, whose results are then considered in aggregate form in the regions of the TIAM model in a following model run. For time and resource reasons, this soft-linking of the different participating models has not been undertaken in the LCS scenario analysis. It has to be noted, also, that a model region (with the exception of the above sub-regions) does not contain further geographical detail. The energy transport infrastructure and its costs within a region can therefore only be represented based on average assumptions about scale and distance. Furthermore, the model approach stipulates cooperative and rational behaviour of the agents in the energy system, i.e. it is assumed that producers, such as OPEC, cannot exercise market power, and that consumers are making purchase decisions for energy appliances based on life-cycle costs, which is contrary to observed behaviour in reality. Consumer behaviour is only partially reflected in TIAM by hurdle rates for investment decisions. Depending on the considered model horizon, input data assumptions have to be made for several decades. The inherent uncertainty contained in this kind of projection requires a careful analysis of the sensitive factors influencing the model results. Due to this data uncertainty, the value of the model analysis lies not so much in the pure numerical results, but in addressing the interactions in the energy system and in deriving results on a more qualitative or relative basis in terms of technology choices.

3. Scenario analysis

In this section, the scenarios analysed with the TIAM model are defined, their results discussed, and some conclusions on technology policy derived.

3.1. Characterization of scenarios

With the TIAM model, three scenarios have been analysed as commonly defined in the LCS project:

1. *Base* scenario without any explicit CO_2 mitigation measures (BASE)
2. *CO_2 price* scenario with CO_2 price rising from \$10/tCO_2 in 2013 to \$100/tCO_2 in 2050 (C10)
3. *CarbonPlus* scenario stipulating a 50% reduction of CO_2 emission in 2050 compared to 2000 (CPLUS).

The BASE scenario describes a development where no initiatives are undertaken to avoid CO_2. It serves, in the scenario analysis, as a benchmark against which to measure the impacts in terms of energy use, technology choice, emissions and costs in the two CO_2 abatement scenarios C10 and CPLUS. The CO_2 price in the C10 scenario does not necessarily have be implemented as a carbon tax, but could represent a bundle of different policies, such as voluntary programmes, subsidies, standards etc., that make low-carbon technologies with abatement costs of up to \$100/tCO_2 in 2050 cost-competitive. While the C10 scenario analyses the CO_2 reduction that can be achieved by abatement measures having mitigation costs of up to \$100/tCO_2, the CPLUS scenario requires (in line with the proposal at the G8 summit in Germany in June 2007) a CO_2 reduction of 50% in 2050 relative to the global emission level of 23.5 GtCO_2 in 2000. In the CPLUS scenario, this reduction target has been formulated as a global target, i.e. depending on the mitigation costs and potentials, some regions can mitigate more than the required quota of 50% and sell their emission reductions to other regions to fulfil their obligations. This resembles a cap-and-trade scheme on a global level, with emission reductions being realized where they can be achieved in the most cost-effective manner. The analysis focuses on the mitigation option for CO_2; other greenhouse gases such as N_2O and CH_4 have been not included in this analysis.

The projected useful energy demand vectors in the agriculture, commercial, residential and transport sectors are derived using demand drivers as described in the previous section. Two important drivers in this context are the development of the population and the gross domestic product (GDP) in the different world regions. In the scenario analysis done with the TIAM model, it has been assumed that the global population will increase from 6.4 billion in 2005 to 9 billion in 2050. Global GDP is assumed to grow at an annual average rate of 2.8% in the period 2000–2050, similar to the B2 scenario group in the IPCC SRES report (IPCC, 2000).

In addition to the three scenarios, two variants of the CPLUS scenarios have been analysed focusing on the contributions of nuclear power (CPLUS-NUC) and carbon capture and storage

(CPLUS-NCCS) to the challenge of reducing CO_2 emissions. In the BASE, C10 and CPLUS scenarios, the share of nuclear power in electricity generation has been restricted to reflect the circumstance that the penetration of nuclear power is not so much restricted for technical or economic reasons but because of issues of public acceptance, nuclear waste storage and non-proliferation. Based on low-growth projections in two studies (MIT, 2003; IAEA, 2007), the maximum nuclear generation has been limited to 8,400 TWh in 2050 in the BASE, C10 and CPLUS scenarios, whereas in the CPLUS-NUC variant a two-thirds higher nuclear power generation of 14,000 TWh can be reached in 2050 (based on a high-growth scenario, as described in MIT, 2003). In the second variant CPLUS-NCSS, the option to capture and store CO_2 in the power and upstream sector has been excluded.

3.2. Scenario results

The discussion of the results starts with the CO_2 emission profiles followed by the developments in electricity generation, final energy use and primary energy consumption, before the cost implications of the scenarios are highlighted. Finally, the results of the sensitivity analysis regarding the role of nuclear power and carbon capture and storage (CCS) are presented.

3.2.1. CO₂ emissions

In the BASE scenario, the global CO_2 emissions will more than double from 23 $GtCO_2$ in 2000 to 48 Gt in 2050 (Figure 2). Compared with the year 2000, the scenario C10 with a CO_2 price of \$100/t in 2050 results in an emission reduction of only 23%; however, relative to the BASE scenario, emissions in 2050 are 60% lower. About 11 Gt of this reduction is achieved by capturing CO_2 at coal-fired (9 Gt) and synfuel production plants (2 Gt) and storing the CO_2 in deep saline aquifers, coal seams, or oil and gas fields. At a sector level, the conversion sector (power and upstream sector) is the only one which reduces its emissions relative to the year 2000 (Figure 2). Combined emissions of the residential, commercial and agriculture sectors are stabilized, while emissions from transportation are still above 2000 levels in 2050.

The required CO_2 mitigation of 50% in the CPLUS scenario in 2050 translates into a 72% reduction compared with the emissions of the BASE scenario in the same year. Overall captured CO_2 emissions

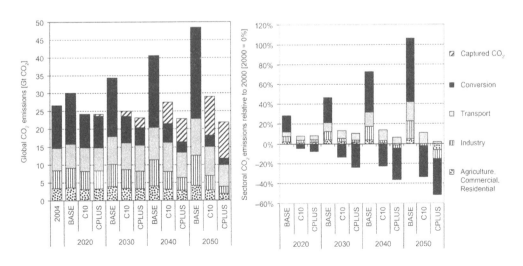

FIGURE 2 Global CO_2 emissions by sector in the BASE, C10 and CPLUS scenarios in absolute figures (left) and relative to the 2000 levels (right).

in the CPLUS scenario amount to 10 Gt which, despite the more stringent CO_2 reduction, is slightly lower than in the C10 scenario, which is explained by the fact that, in contrast to the C10 scenario, the majority of CO_2 capture takes place at gas rather than coal power plants. Because of the lower carbon content of natural gas compared with coal, this results in lower quantities of CO_2 needing to be stored. The contribution to attaining the 50% reduction target differs between the sectors.

The major share of the reductions relative to 2000 is achieved in the conversion sector, accompanied by reduction efforts of the industry and building sectors, whereas transport sector emissions are stabilized in the CPLUS scenario at their 2000 level (Figure 2).

Economic uptake in developing countries at an assumed average growth rate of 4.4%/year is accompanied by a tripling of CO_2 emissions in the BASE scenario between 2000 and 2050. In 2050, 52% of global CO_2 emissions originate from developing countries (Figure 3). In the C10 scenario, OECD regions reduce their emissions by 44% relative to the corresponding level in 2000, whereas emissions in developing countries are 19% higher than in 2000, but are still lowered by two-thirds compared with the BASE level. To go from the C10 scenario with 23% lower global CO_2 emissions in 2050 to the 50% reduction goal of the CPLUS scenario, emissions have to be reduced by an additional 6.5 Gt, of which 3.1 Gt is mitigated in the developing countries and 3.4 Gt combined in the OECD and transition economies. Relative to the year 2000, this corresponds to a reduction of 21% for the developing countries, 50% for the transition economies, and 63% for the OECD. For the Western Europe region,[3] the resulting CO_2 emission pathway in the CPLUS scenario is with reductions relative to the Kyoto reference year 1990 of 16% by 2020, 30% by 2030, and 70% by 2050 in the range of reduction targets being discussed, with –60% in the UK and –80% in Germany by 2050.

On a 'per capita' basis, the average global per capita CO_2 emissions decline from 3.7 t/cap in 2000 to 1.3 t/cap in the CPLUS scenario, but regional per capita emissions differ significantly in 2050: from 3.4–4.4 t/cap in the OECD or transition economies to 0.8 t/cap in the developing countries. This is due to the fact that the carbon mitigation takes place in the CPLUS scenario in those regions where it is most cost-effective, allowing for full trade in emission permits and unhindered technology transfer between the world regions. Therefore, the latter discussed mitigation costs represent a lower bound on the costs, restrictions or barriers to the adoption of these flexible instruments (e.g. limitations in the permit trade), which will result in an increase in the abatement costs.

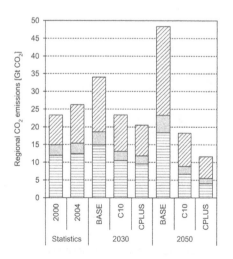
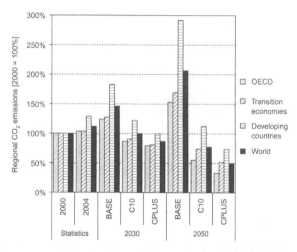

FIGURE 3 Regional breakdown of CO_2 emissions in absolute figures (left) and relative to the year 2000 (right).[4]

3.2.2. Electricity generation

Based on a global fuel demand of 173 EJ in 2004, power generation is a major energy consumer, being responsible for 37% of the total primary energy consumption in 2004 and 11 Gt or 41% of global energy-related CO_2 emissions. In the BASE scenario, electricity generation nearly triples between 2004 and 2050 (Figure 4), with 50% of the electricity being produced in developing countries compared to 33% in 2004. Lower fuel prices compared with natural gas, and an absence of CO_2 mitigation policies, strengthen the position of coal in the BASE scenario as the major fuel for electricity generation: 62% or 28,950 TWh of electricity is produced in coal plants in 2050. Nuclear power plants with an electricity generation cost range of $21–69/MWh would be competitive with coal plants with a range of $16–69/MWh (as reported in NEA/IEA, 2005, for different countries and discount rates); but, as already discussed, because of a lack of public acceptance, their generation has been limited to 8,400 TWh in the scenarios. The results of a sensitivity analysis allowing for a greater share of nuclear based generation are discussed later.

Electricity generation in the C10 scenario is characterized, on the one hand, by a drop in the total production compared to the BASE scenario (–17% in 2050) due to efficiency and saving efforts in the end-use sectors and, on the other hand, by an increased generation from wind and biomass and natural gas at the expense of coal. After 2030, the initial increase in natural gas usage is reversed through the increased amounts of coal being burnt in power plants fitted with a mechanism for the capture and subsequent storage of CO_2 (9 Gt in 2050). Due to the changed electricity mix, the CO_2 intensity of power generation drops from 460 g/kWh in the BASE scenario to 45 g/kWh in the C10 scenario in 2050.

In the CPLUS scenario, the decarbonizing of the power sector is continued, yielding a CO_2 intensity of 23 g/kWh in 2050. This low carbon intensity of electricity, achieved by a switch from coal to gas plants with CO_2 capture, allows the increased use of electricity in the end-use sectors to substitute their fossil energy carriers. However, this mitigation strategy requires that power generation is almost entirely decarbonized. Also, the change from coal to natural gas carbon capture plants

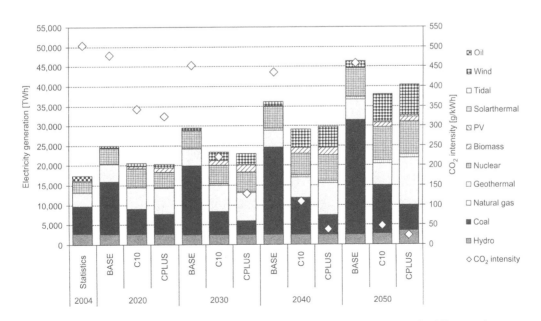

FIGURE 4 Electricity generation by fuel and CO_2 intensity in the BASE, C10 and CPLUS scenarios.

is economically viable only at CO_2 prices beyond $100/tCO_2$. This choice is, however, sensitive to the relationship of the coal and gas price as well as the assumed techno-economic characteristics of the capture plants. Generation from renewables is only slightly higher in the CPLUS scenario compared with the C10 scenario, since carbon capture and storage is the more cost-effective mitigation option under the assumed conditions. The relationship between CCS and renewables in power generation is analysed in a sensitivity analysis in Section 3.2.6.

3.2.3. End-use sectors

In the BASE scenario, the total final energy demand for the agriculture, residential, commercial, industry and transport end-use sectors grows by 73% between 2004 and 2050 (Figure 5). The industry and transport sectors are the two major final energy consumers, with shares of 39% and 27%, respectively, in total final energy use in 2050. With a consumption of 149 EJ (29%) in 2050, electricity has a similar market share as petroleum (27%). Nearly half of the electricity is consumed in the industry sector for heating and machine-drive services, while petroleum still covers 92% of the transport sector's energy demand. Despite the absolute increase in total final energy consumption in the BASE scenario, the overall energy efficiency of the end-use sectors, defined here as total final energy use per GDP, is decreasing at an annual average rate of 1.5%.

In the C10 scenario, the CO_2 penalty triggers an 11% decline of final energy demand compared with the BASE scenario in 2050, or an annual efficiency improvement of 1.8% relative to 2000. Particularly responsible for this reduction are efficiency improvements and savings in the industry sector (−16% relative to the BASE scenario in 2050) through more efficient motors and boilers, and a switch to cogeneration of process heat and electricity. In the residential and commercial sectors, better building insulation, more efficient electric appliances, a growth in district heat from CHP plants, and efficiency gains for room heating, cooling and hot water preparation, e.g. by the use of dual-mode heat pumps (reaching a coefficient of performance of up to 5 for heating and 6–7 for cooling), contribute to the reduction in final energy demand (−12% relative to the BASE scenario).

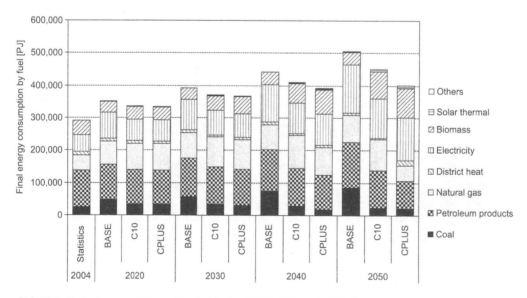

FIGURE 5 Final energy demand by fuel in the BASE, C10 and CPLUS scenarios.

The fuel mix in the C10 scenario is characterized by a switch to biomass, driven by a substitution of petroleum by ethanol or DME in the transport sector. This trend towards electricity continues in the CPLUS scenario, where electricity consumption in 2050 is 6% higher than in the C10 scenario.

While natural gas consumption in the C10 scenario is increasing (+13%) compared with the BASE case, as a substitute for coal, it is drastically reduced by 50% or 49 EJ in the CPLUS scenario, particularly through the use of heat pumps and industrial biomass combined heat and power (CHP) plants. These saved gas amounts are nearly equivalent to the increased use of gas in power generation in the CPLUS scenario, so that the total primary consumption of natural gas between the C10 and CPLUS scenarios remains almost unchanged. In addition, in the CPLUS scenario, hydrogen is produced from coal gasification plants with CO_2 capture, giving hydrogen a 5% share of transport fuel demand in 2050. Overall final energy consumption in the CPLUS scenario in 2050 is 20% below the BASE scenario, corresponding to an efficiency improvement of 2.1%/year. The CO_2 intensity of final energy use falls from 28 kg CO_2/GJ in the BASE run to 19 kg CO_2/GJ in the CPLUS scenario in 2050.[5] Mainly responsible for this intensity decrease are the substitution of fossil fuels by biomass, district heat and electricity, with the latter two being produced with very low CO_2 emissions in the power sector.

3.2.4. Total primary energy supply

The resulting total primary energy supply (TPES) caused by the developments in the conversion and end-use sectors is shown in Figure 6. In the BASE scenario, the combined share of the fossil fuels – coal, gas and oil – remains almost constant at about 80% between 2004 and 2050, but the share of coal in TPES nearly doubles in relative figures from 20% to 44%, and triples in absolute figures in the same time due to the consumption in power generation and the production of FT fuels. In the mitigation scenarios C10 and CPLUS, the share of fossil fuels decreases to 64% and 56%, respectively, in 2050, whereas renewable sources contribute 23% and 30%, respectively, to

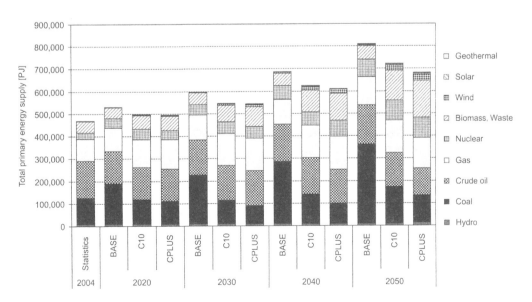

FIGURE 6 Total primary energy supply by fuel in the BASE, C10 and CPLUS scenarios.[6]

the primary energy consumption. Biomass becomes the single largest energy carrier in the CPLUS scenario in 2050, with a consumption of 155 EJ, of which 60 EJ is used for the production of transport fuels, 70 EJ as combustion fuel in the other end-use sectors, 15 EJ in power and heat plants, and 10 EJ as feedstock in industry. For the global biomass potential an amount of 550 EJ, based on Smeets et al. (2007), has been assumed in the analysis.

3.2.5. Cost effects

The cost implications of the mitigation scenarios are summarized in Table 1. The difference in the annual energy system costs displayed for the years 2030 and 2050 describes the additional global costs of the energy system in the mitigation scenarios C10 and CPLUS in real terms relative to the BASE scenario in the same year. It needs to be pointed out that this cost assessment only covers the energy sector; impacts on other industry sectors affecting the development of the entire economy in the form of GDP or employment rate are not captured in these cost figures. Through these interactions the energy demand, and thus the cost impacts directly occurring in the energy sector, can be reduced, although this may be offset by additional costs in other parts of the economy.

In the C10 scenario, the stabilization of global CO_2 emissions in 2030 at levels comparable to those in 2000, yields additional annual costs of $13 billion. In later years the tripling of the CO_2 price from $29 to $100/$tCO_2$ in 2050 is reflected by a drastic increase in the annual costs, culminating in additional costs of $641 billion in 2050 to reduce CO_2 emissions by 23%. To realize the 50% reduction target (roughly a doubling of the reduction efforts relative to C10) in the CPLUS scenario, the additional annual costs in 2050 of $1,633 billion are nearly three times higher than the cost difference of the C10 scenario. These additional costs are equivalent to 1.3% of the assumed global GDP of $126 trillion in 2050.

In the C10 scenario, all CO_2 abatement options are realized with marginal mitigation costs[7] being less than or equal to $100/$tCO_2$. To achieve the stipulated CO_2 reduction of 50% in the CPLUS scenario, additional abatement measures with higher costs have to be utilized, leading to

TABLE 1 Cost indicators for reduction scenarios C10 and CPLUS

Indicator	Region	Units	Scenario			
			C10		CPLUS	
			2030	2050	2030	2050
Difference annual	OECD	Billion $\$_{2000}$	8	332	96	727
energy system	Transition	Billion $\$_{2000}$	1	79	21	195
costs relative to	economies					
BASE scenario	Developing countries	Billion $\$_{2000}$	4	230	10	711
	World	Billion $\$_{2000}$	13	641	126	1633
Marginal abatement costs	World	$\$_{2000}$/$tCO_2$	29	100	63	330
Average abatement costs	World	$\$_{2000}$/$tCO_2$	2	21	10	44

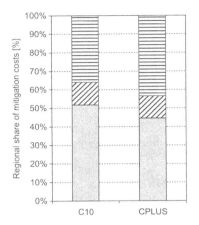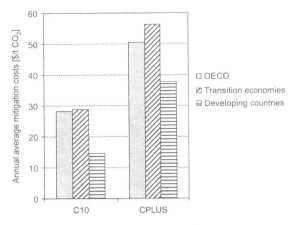

FIGURE 7 Regional breakdown of mitigation costs of C10 and CPLUS scenarios (left) and regional average mitigation costs (right) in 2050.

marginal abatement costs for the 50% reduction target of $330/tCO_2$. The marginal abatement costs are only an indicator for the mitigation costs of the last tonne of CO_2 which has to be avoided in order to achieve the reduction target. Although these marginal costs can be quite high, the average mitigation costs, which are calculated by dividing the cost differences from the BASE scenario shown in Table 1 by the corresponding CO_2 reductions, are typically much lower, since low-cost abatement measures, being realized earlier on the reduction path, are also being considered in this cost indicator. In the C10 scenario, average mitigation costs in 2050 are $21/tCO_2$, while in the CPLUS scenario average mitigation costs are, at $44/tCO_2$, about twice as high.

On a regional level, half of the mitigation costs in the C10 scenario accrue in OECD countries (Figure 7); increasing CO_2 abatement targets in the CPLUS scenario moves the majority of these costs to non-OECD countries, which have to carry 60% of the annual cost difference relative to the BASE scenario in 2050. However, owing to the higher absolute emissions reductions achieved with these funds in the developing countries, average mitigation costs in these countries in the C10 and CPLUS scenarios are still lower than in the OECD and transition economies.

3.2.6. Sensitivity analysis regarding CCS and nuclear power

In a sensitivity analysis for the 50% reduction scenario CPLUS, the roles of CCS and nuclear power were analysed. In the scenario variant CPLUS-NCCS, the option of capturing and storing CO_2 is not available, while in the variant CPLUS-NUC the maximum allowed nuclear power production in 2050 is increased from 8,400 to 14,000 TWh.

If CCS is not available to make a contribution to reaching the 50% CO_2 mitigation target in 2050, renewable energy carriers will play a more important role in power generation. The renewable share increases from 34% in the CPLUS scenario to 72% in the CPLUS-NCCS variant. In addition, the increased electricity generation costs due to the more expensive generation from renewable sources lead to a reduced electricity demand in the end-use sectors. This reduction is partially achieved by electricity-saving measures, but also by a higher natural gas consumption, which again triggers efficiency improvements in the transport sector to offset the increment in CO_2 emissions from natural gas. Overall, the missing CCS option yields (with $220/tCO_2$) five times higher average abatement costs in 2050 compared with the CPLUS scenario.

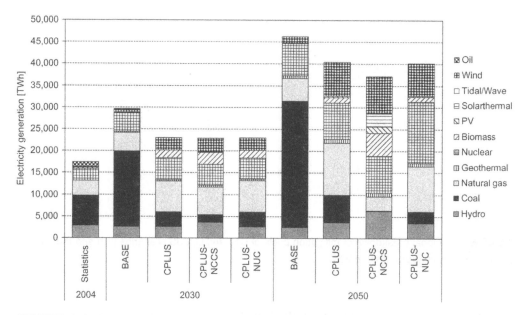

FIGURE 8 Electricity generation in the variants of the CPLUS scenario.

In the variant CPLUS-NUC the additional electricity generation from nuclear (compared to the scenario CPLUS) replaces base- to mid-load generation from coal and natural gas plants that are equipped with carbon capture mechanisms (Figure 8). Thus, the amounts of captured and stored CO_2 are reduced from 11 to 7 Gt in 2050. The overall electricity generation remains nearly the same, so that the effects of increased nuclear power generation are mainly limited to the power sector. The two-thirds higher nuclear generation in 2050 yields a 9% reduction in the annual mitigation costs compared with those for the CPLUS scenario shown in Table 1.

Overall the sensitivity analysis shows that the penetration level of renewable technologies in the electricity sector depends, as well as on the technology characteristics themselves, also on the availability or success of competing options such as CCS. If CCS becomes available on a commercial scale by 2015–2020, it may reduce the need for renewable technologies as long as sufficient storage volumes exist. In a similar way, nuclear reduces the need for CCS. This ranking depends, however, on the future development of the different technologies, which is uncertain, especially in the case of renewables and CCS, so that a technology strategy should build on a mix of technologies in order to hedge against these uncertainties.

3.3. Technology and policy implications

Without explicit policies in place to avoid carbon emissions, as in the BASE scenario, the development of the global energy system is characterized by an increased reliance on fossil fuels, especially coal in power generation and for the production of synthetic fuels. Despite 'no-regrets' efficiency improvements in the conversion and end-use sectors, this trend to carbon-intensive fuels leads to a doubling of global CO_2 emissions from 2000 to 2050.

To substantially reduce CO_2 emissions, as envisaged in the mitigation scenarios, no single mitigation option can achieve the required reductions on its own; but instead a portfolio of technology options and measures is required. The power sector, being the largest emitter in the BASE scenario, is the key to bring down CO_2 emissions. In the C10 scenario, CCS in electricity generation is responsible for emission reductions of up to 9 Gt in 2050. In addition, cogeneration of heat and electricity, especially in industry, represents (due to its high overall efficiency) a further option for emission reductions. Renewable electricity options such as wind converters, biomass CHPs or geothermal plants are generally not yet cost-competitive, but will start to gain market shares at CO_2 prices of $10–20/tCO$_2$, as observed for example in the C10 scenario after 2020. Nuclear power represents a further abatement option in the scenario analysis, but public acceptance is critical for its future role.

Because of its relatively low reduction costs and high abatement potential, the power sector is an important element in both reduction scenarios. However, with increasing CO_2 price, mitigation options in the end-use sectors are also becoming more and more important. Reduction efforts that are effective in the industry sector are in the C10 scenario efficiency and saving measures, e.g. by better process integration to recover heat. In the building sector, efficiency standards to reduce the heating and cooling demands of new and existing buildings already exist in many countries and have been proven to be cost-effective. In the transport sector, bio-ethanol is introduced at CO_2 prices of $20–30/tCO$_2$ in the C10 scenario; and hydrogen fuel-cell cars enter the market at prices above $90/tCO$_2$. At CO_2 prices beyond $60/t, the initial trend from coal to natural gas in the end-use sectors is altered by substituting, where possible, natural gas by electricity, e.g. compression heat pumps in the residential sector or plug-in hybrid cars in the transport sector. However, a prerequisite for this electrification strategy, in order to reduce the overall emissions of the energy system, is a nearly completely decarbonized power sector.

The scenario and model analysis is based on an optimistic and idealized view of the future energy sector. It is optimistic in a sense that it is assumed that policies, e.g. in the form of research, development and deployment (RD&D), are in place to solve technical problems in the development of as yet immature technologies such as CCS or to support the deployment phase of new technologies through instruments such as financial incentives or regulations. It is idealized since resistance to the introduction of mitigation measures is generally not considered in the model analysis, e.g. lack of information about the existence of efficient appliances. In the scenario it is stipulated that such barriers have been removed by adequate policy instruments such as, in the case of lack of information, by information campaigns or best-practice examples to raise awareness of more efficient technologies. A further barrier is represented by the absence of regulation or standards in new technology areas such as for the build-up of a hydrogen infrastructure, the blending of biofuels with regular transport fuels, or concerning the licensing, liability and leakage aspects of CO_2 storage sites. Uncertainty regarding long-term climate policies might lead to a postponement of investment decisions or, at worst, lead to lock-in into carbon-intensive technologies, thus jeopardizing the achievement of deep reduction targets later.

The majority of the CO_2 will be emitted in developing countries in 2050, as indicated in the BASE scenario. Cooperation between developed and developing countries in terms of technology transfer and financing becomes essential in order to reduce global emissions significantly. The comparison of the regional mitigation costs in the C10 and CPLUS scenarios shows that the average mitigation costs can be up to 50% lower in developing countries compared with OECD countries.

4. Conclusions

Using the global ETSAP-TIAM, model scenarios to achieve deep emission reductions up to a level of 50%, as proposed by the G8 in Heiligendamm 2007, were analysed. Without any reduction obligation, global CO_2 emissions in the BASE scenario more than double globally between 2000 and 2050. Strategies to reverse this unsustainable pathway were studied in two mitigation scenarios. The C10 scenario, stipulating a common CO_2 price in all world regions increasing from $10/tCO$_2$ in 2013 to $100/t in 2050, yields a global reduction of 23% in 2050 relative to the year 2000. A doubling of these mitigation efforts in order to reach the global target of a 50% reduction was analysed in a second abatement scenario (CPLUS), resulting in marginal abatement costs of $330/tCO$_2$ in 2050. The regional mitigation obligations differ between world regions. OECD regions such as Western Europe must reduce their emissions by 70% relative to 1990, which lies in the range of reduction targets proposed by some European countries (e.g. –60% by the UK), whereas developing countries need to achieve, on average, a reduction of 23% by 2050. As central mitigation options have been identified in the conversion sector, the use of coal or gas power plants with CO_2 capture and the increased use of renewable energy sources, namely wind, biomass and geothermal. In the end-use sectors, efficiency improvements in industry (motors, boilers) as well as the building sector (heating, hot water boilers), the use of biofuels in transport and heat pumps in the building sector are common elements in both mitigation scenarios. The role of natural gas as a final energy resource is sensitive to the CO_2 price or the reduction target: moderate reduction efforts as in the C10 scenario favour the switch from coal to natural gas, especially in industry, whereas for deeper emission reductions, as in CPLUS scenario, natural gas is replaced by electricity being produced with low specific emissions in the power sector, which also requires further reduction efforts in the power sector by switching from coal- to gas-fired power plants with CO_2 capture. The prerequisites for technologies to play this role in achieving a low-carbon society are increased RD&D efforts to bring new low-carbon technologies to the market, measures to overcome non-economic barriers, and the establishment of long-term policy instruments stimulating the investment in carbon abatement. The focus of the analysis made here with the TIAM model has been on the detailed technological assessment of measures that need to be taken in order to achieve mitigation targets; whereas the quantification of economic impacts of climate policies (economic growth, population, trade flows) lies beyond this model framework and will need to be addressed with economic models.

Notes

1. The Energy Technology Systems Analysis Programme (ETSAP) is an implementing agreement of the International Energy Agency (IEA), first established in 1976. It functions as a consortium of member country teams and invited teams that actively cooperate to establish, maintain and expand a consistent multi-country energy/economy/environment/engineering (4E) analytical capability. See www.etsap.org/index.asp for further information.
2. Individual countries represented in TIAM are Canada, China, India, Japan, Mexico, South Korea and the USA.
3. Western Europe includes EU-15 plus Iceland, Greenland, Malta, Norway and Switzerland.
4. The group of developing countries includes Africa, Central and South America, the Middle East, China, India and other Asian countries (except South Korea and Japan). For historical reasons, the OECD region does not contain Hungary, Czech Republic and Slovak Republic. These countries are still part of the group of transition economies, whose other members include the succession states of the Former Soviet Union, Albania, Bosnia-Herzegovina, Bulgaria, Croatia, Serbia and Montenegro, the former Yugoslav Republic of Macedonia, Latvia, Lithuania, Romania and Slovenia.
5. For comparison, natural gas has a CO_2 intensity of 56 kg CO_2/GJ, and hard coal a CO_2 intensity of 94 kg CO_2/GJ.
6. Based on a direct equivalent method assuming an efficiency of 100% for the use of solar and wind energy in power generation.
7. Marginal abatement costs are defined as the costs related to the abatement of one additional tonne of CO_2 from a given reduction target.

References

EC (Council of the European Union), 2007, *Presidency Conclusions*, 8–9 March 2007 [available at www.consilium.europa.eu/ueDocs/cms_Data/docs/pressData/en/ec/93135.pdf].

ETSAP (Energy Technology Systems Analysis Programme), 2007, *Models and Applications: Global* [available at www.etsap.org/applicationGlobal.asp].

IAEA (International Atomic Energy Agency), 2007, *Energy, Electricity and Nuclear Power Estimates for the Period up to the Year 2030*, Reference Data Series No.1, IAEA, Vienna.

IEA (International Energy Agency), 2006, *CO_2 Emissions from Fossil Fuel Combustion 1971–2004*, 2006 Edition, IEA, Paris.

IPCC, 2000, *Emissions Scenarios: Special Report of the Intergovernmental Panel on Climate Change*, N. Nakicenovic, R. Swart (eds), Cambridge University Press, Cambridge, UK.

Loulou, R., Labriet, M., 2007a, 'ETSAP-TIAM: the TIMES integrated assessment model. Part I: model structure', *Computational Management Science*, doi:10.1007/s10287-007-0046-z.

Loulou, R., Labriet, M., 2007b, 'ETSAP-TIAM: the TIMES integrated assessment model. Part II: mathematical formulation', *Computational Management Science*, doi:10.1007/s10287-007-0045-0.

Loulou, R., Remme, U., Kanudia, A., Lehtilä, A., Goldstein, G., 2005, *Documentation of the TIMES Model* [available at www.etsap.org/documentation.asp].

MIT (Massachusetts Institute of Technology), 2003, *The Future of Nuclear Power: An Interdisciplinary MIT Study*, MIT Press, Cambridge, MA.

NEA (Nuclear Energy Agency), IEA (International Energy Agency), 2005, *Projected Costs of Generating Electricity: 2005 Update*, OECD Publishing, Paris.

Remme, U., Blesl, M., Fahl, U., 2007, *Global Resources and Energy Trade: An Overview for Coal, Natural Gas, Oil and Uranium*, Research Report 101, Institute of Energy Economics and the Rational Use of Energy, University of Stuttgart [available at http://elib.uni-stuttgart.de/opus/frontdoor.php].

Smeets, E.M.W., Faaij, A.P., Lewandowski, I.M., Turkenburg, W.C., 2007, 'A bottom-up assessment and review of global bio-energy potentials to 2050', *Progress in Energy and Combustion Science* 33, 56–106.

Ürge-Vorsatz, D., Harvey, L.D.D., Mirasgedis, S., Levine, M.D., 2007a, 'Mitigating CO_2 emissions from energy use in the world's buildings', *Building Research and Information* 35(4), 379–398 (doi:10.1080/09613210701325883).

Ürge-Vorsatz, D., Koeppel, S., Mirasgedis, S., 2007b, 'Appraisal of policy instruments for reducing buildings' CO_2 emissions', *Building Research and Information* 35(4), 458–477 (doi:10.1080/09613210701327384).

Yohe, G.W., Lasco, R.D., Ahmad, Q.K., Arnell, N.W., Cohen, S.J., Hope, C., Janetos, A.C., Perez, R.T., 2007, 'Perspectives on climate change and sustainability', in: M.L. Parry, O.F. Canziani, J.P. Palutikof, P.J. van der Linden, C.E. Hanson (eds), *Climate Change 2007: Impacts, Adaptation and Vulnerability*. Contribution of Working Group II to the Fourth Assessment Report of the Intergovernmental Panel on Climate Change, Cambridge University Press, Cambridge, UK, 811–841.

■ research article

Implications for the USA of stabilization of radiative forcing at 3.4 W/m²

JAE EDMONDS*, LEON CLARKE, MARSHALL WISE, HUGH PITCHER, STEVE SMITH

Joint Global Change Research Institute, 8400 Baltimore Avenue, Suite 201, College Park, MD 20740-2496, USA

Stabilization presents a daunting challenge for all countries of the world, regardless of their stage of development, institutions or technological capabilities. This article explores the implications for the USA of climate stabilization at 3.4 W/m². Stabilization at this level, even under idealized conditions of nearly immediate global cooperation, will require a transformation of the USA's energy system, beginning almost immediately and extending throughout the century and beyond. This transformation will need to be even more rapid and extensive if the emissions reduction regime encompasses only a portion of the global economy. The availability of advanced technologies such as CCS, sustainable bioenergy production, wind and solar, nuclear energy and end-use efficiency improvements will facilitate this transition. Indeed, the degree to which technology advances over the coming century is among the most important determinants of the economic costs of stabilization for the USA and the rest of the world. The scope of the energy system transformation highlights the need to begin deploying technologies that are currently available and to continue to invest in R&D to develop newer, more efficient, and less expensive low- or zero-carbon energy supply technologies and end-use technologies.

Keywords: climate change; climate stabilization; energy; energy systems; low-carbon society; technology

La stabilisation représente un défi grave pour tous les pays du monde, quelque soit leur stade de développement, leurs institutions ou aptitudes technologiques. Ce papier explore les conséquences éventuelles pour les Etats-Unis d'une stabilisation climatique à 3.4 W/m². Une stabilisation à ce niveau, même dans des conditions idéalisées de coopération mondiale presque immédate, nécessitera une transformation du système énergétique des Etats-Unis à commencer dès à présent et se déroulant tout au long de ce siècle et au-delà. Cette transformation devra être d'autant plus rapide et extensive si le régime de réduction des émissions n'englobe qu'une partie de l'économie mondiale. La disponibilité de technologies avancées telles que la CSC, la production durable de bioénergie, l'éolien et le solaire, l'énergie nucléaire et l'amélioration de l'efficacité à utilisation finale faciliteront cette transition. En effet, l'avancée des progrès technologiques au cours du siècle à venir sera un des facteurs les plus déterminants du coût économique de stabilisation pour les Etats-Unis et le monde. L'etendue de la transformation du système énergétique met en valeur le besoin d'amorcer le déploiement des technologies qui sont disponibles actuellement et de continuer à investir dans la recherche et le développement pour le développement de technologies d'énergie faiblement carbonées ou décarbonées et de technologies d'utilisation finale plus neuves, plus efficaces, moins chères.

Mots clés: changement climatique; énergie; société sobre en carbone; stabilisation du climat; systèmes énergétiques; technologie

■ *Corresponding author. E-mail: jae@pnl.gov

CLIMATE POLICY 8 (2008) S76–S92

doi:10.3763/cpol.2007.0495 © 2008 Earthscan ISSN: 1469-3062 (print), 1752-7457 (online) www.climatepolicy.com

1. Introduction

Stabilization is a global challenge: no country can stabilize atmospheric GHG concentrations acting alone. In the long run, stabilization will require the combined efforts of nearly every country around the world. But every country and region is different; in its history, its demographics, its institutions, its economic prosperity, its energy system, its technological capacity and, of course, its GHG emissions. The challenges presented by stabilization in every country are therefore different, but they are also linked because all countries must eventually participate in emissions mitigation.

Limiting anthropogenic climate change requires the stabilization of radiative forcing, a measure of the change in atmospheric energy balance by greenhouse gases (GHGs) and aerosols.[1] This article explores the potential implications of stabilization for the USA within a global context. We explore the US implications within a cooperative global scenario in which radiative forcing from carbon dioxide (CO_2), methane (CH_4), nitrous oxide (N_2O), hydrofluorocarbons (HFCs), perfluorocarbons (PFCs) and sulphur hexafluoride (SF_6) is limited to 3.4 W/m², relative to a pre-industrial state. This radiative forcing level is consistent with stabilizing the concentration of CO_2, the most important GHG released by humans to the atmosphere, at approximately 450 parts per million (ppm), stabilizing the concentration of CH_4 at approximately 1.4 ppm, and stabilizing the concentration of N_2O at 0.36 ppm. There is no scientific consensus that limiting radiative forcing to 3.4 W/m² is the 'right' target. Nevertheless, limits in this range are of particular interest from the perspective of the low-carbon society and are the subject of this series of papers in this *Climate Policy* supplement.

This article illustrates that stabilization will require substantial changes in the energy sector, changes that will ultimately encompass every country either directly or indirectly. The scenario presented here envisions a world where increasing prosperity results in the majority of global emissions originating from non-Annex 1 countries by 2020. The US economy grows fivefold over the century, leading to emissions increases, without mitigation actions, of only 35%, in large part because of substantial advances in energy supply and demand technologies. Even in an optimal global strategy, stabilization at 3.4 W/m² still requires a transformation of the US energy system that needs to begin almost immediately. Delays in emissions reductions by non-Annex 1 countries increase the speed and extent of this transformation (Edmonds et al., 2008).

2. The MiniCAM model

The analysis presented here employs the ObjECTS MiniCAM model (Brenkert et al., 2003; Kim et al., 2006). The ObjECTS MiniCAM is a long-term, global integrated assessment model of energy, economy, agriculture and land use, which considers the sources of emissions of a suite of greenhouse gases (GHGs) emitted in 14 globally disaggregated regions, the fate of emissions to the atmosphere, and the consequences of changing concentrations of greenhouse-related gases for climate change over a time period ranging from 1990 to 2095. The model combines a technologically detailed global energy-economy model, an agricultural land-use model (Gillingham et al., 2007), and a suite of coupled gas-cycle, climate, and ice-melt models: the Model for the Assessment of Greenhouse-gas Induced Climate Change (MAGICC; Wigley and Raper, 1992, 2002; Raper et al., 1996). The MiniCAM is a direct descendent of the energy-sector model described by Edmonds and Reilly (1985). MiniCAM has been used extensively for energy, climate, and other environmental analyses conducted for organizations that include the US Department of Energy (DOE), the US Environmental Protection Agency, the IPCC, and several major private sector energy companies. The model is designed to examine long-term, large-scale changes in global and regional energy, economy, emissions of greenhouse gases, short-lived species, and land-cover, atmosphere, carbon cycle, ocean and climate systems, with special emphasis on the role of energy technology.[2]

3. The reference scenario

The reference scenario serves as a point of departure for exploring the implications for the USA of stabilization of global radiative forcing at a level consistent with a 'low-carbon society'. The reference scenario used in this article is described in detail by Clarke et al. (2007a). It assumes that climate policies that are presently in place throughout the world remain in place until the year 2012, at which time they are assumed to expire and are not replaced or extended by other policies motivated by climate change. Other policies and measures motivated by local and regional environmental quality considerations are assumed to be strengthened and extended. For example, sulphur emissions are assumed to be increasingly limited throughout the world. The assumption that no country takes action on climate for the full century is deliberately unrealistic.[3] The reference scenario is constructed as a contrast to alternative scenarios that limit radiative forcing to specific levels (Clarke et al., 2007a).

The most important assumptions shaping the reference scenario are population and labour productivity growth. We assume both a demographic transition and rapid economic expansion that gradually permeates the presently developing world. A central characteristic of the reference scenario is the increasing importance of the non-Annex I nations. World population is increasingly dominated by these countries (Figure 1). Rates of economic growth well above those in the Annex 1 countries also shift economic output to the non-Annex 1 countries (Figure 2). As a result, the non-Annex 1 countries produce more CO_2 than the Annex 1 countries by 2020 (Figure 3). In total, considering all the GHGs in this study, radiative forcing by the end of the century in the reference scenario well exceeds the stabilization limit of 3.4 W/m² considered in this article, and CO_2 takes on an increasing share of the total forcing (Figure 4A).[4]

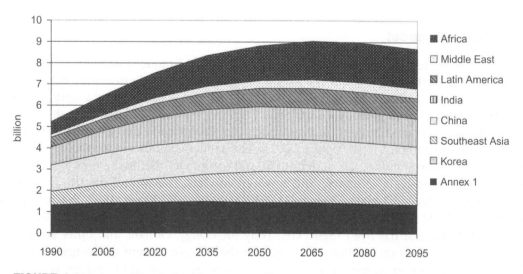

FIGURE 1 Global population in the reference scenario.

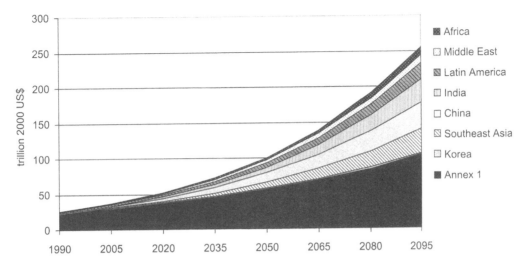

FIGURE 2 Global GDP in the reference scenario.

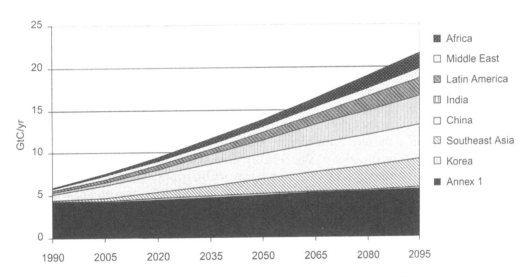

FIGURE 3 Global fossil fuel and industrial carbon emissions by region 2005–2095.

The reference scenario incorporates substantial technological developments and assumes the availability of multiple energy forms. The economic cost and performance of wind and solar power systems are assumed to improve over the course of the century. Nuclear power is assumed to be available and to compete on the basis of cost and performance, and is assumed to successfully address non-economic issues which include nuclear waste, weapons proliferation, energy security, health and safety. Several bioenergy technologies are assumed to be available, including traditional bioenergy fuel use in developing countries, bioenergy based on use of waste and residue products such as bark in the pulp and paper industry, agricultural residues, and dedicated bioenergy crops.

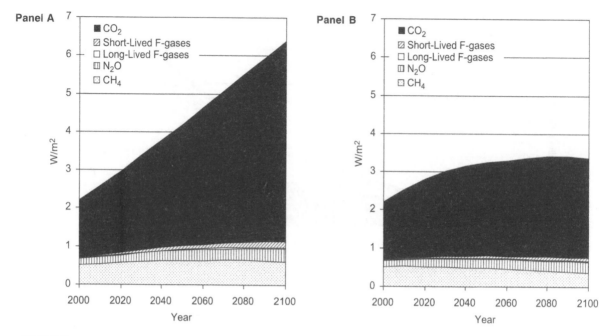

FIGURE 4 Reference case radiative forcing by gas (A) and radiative forcing in the 3.4 W/m² stabilization scenario (B).

Despite the fact that dramatic increases in the production and use of non-fossil energy forms takes place in the reference scenario, the global and US energy systems continue to be dominated by fossil fuel use (Figure 5).[5] In addition, because of the availability of these options, along with assumptions of substantial improvements in energy intensity over the century, US energy primary energy consumption increases by only 35% over the century despite a more than fivefold increase in US economic output. Indeed, the USA's primary energy consumption is level or declining over the final four decades of the century. In a sense, the assumptions underlying this reference scenario lead to emissions reductions, even without policy, beyond what would occur if technology were assumed to remain static or advance more slowly over the century. Emissions would be larger if other forces that influence GHG emissions, such as population growth and per-capita energy service demand growth, were to increase at faster rates than assumed in this scenario.

4. Emissions and stabilization of radiative forcing at 3.4 W/m²

The stabilization scenario assumes that all the world's countries begin, after 2012, to work cooperatively to reduce greenhouse gas emissions. Global economic efficiency is assumed, meaning that emissions reductions are undertaken so as to equalize marginal costs of emissions reductions across regions and GHGs. In addition, the intertemporal allocation of emissions reductions is designed to minimize global costs.

Note that the radiative forcing target here refers to total greenhouse gas forcing only. Forcing from aerosols, tropospheric and stratospheric ozone were not included in the target. The sum of these additional forcings is small by 2100, with positive tropospheric ozone forcing nearly cancelling, on a global average basis, the net negative aerosol forcing, changing the total forcing target by only a small amount.

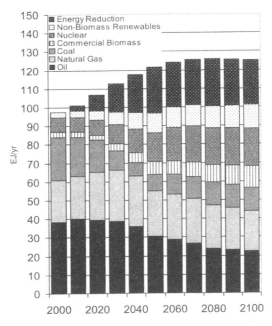

FIGURE 5 The USA's primary energy consumption.

Stabilization of radiative forcing is accomplished by placing an economic value on GHG emissions. In MiniCAM this is accomplished by applying a tax on emissions of GHGs. This tax is uniform across all sources of emissions – industrial, fossil fuel use, and land-use change emissions – and across all regions of the world. Initially, the uniform tax on GHG emissions is assumed to rise at the rate of interest plus the natural rate of removal from the atmosphere (Hotelling, 1931; Peck and Wan, 1996). For example, if the rate of interest is 4% per year and the rate of removal of carbon from the atmosphere is 1% per year, then the price of carbon rises at 5% per year. At the point in time when the concentration of CO_2 reaches the stabilization limit, the price of carbon is set by the physical limit on carbon uptake by natural systems at the steady-state concentration. The time of radiative forcing by gas for the 3.4 W/m² stabilization scenario is shown in Figure 4B.

The price of carbon rises exponentially until mid-century, when it reaches approximately \$500/tC (\$136/tCO$_2$), at which point the concentration of CO_2 approaches its steady-state value (Figure 6). After 2050 the price continues to rise until approximately 2080, but at a slower than exponential rate.[6] The absolute price actually begins to fall at the end of the 21st century as emissions are controlled to maintain the CO_2 concentration at its steady-state value.

Other policy instruments could be employed to achieve the same outcome. However, to minimize cost, an instrument must maintain marginal GHG emissions mitigation costs nearly equal across all regions and emissions sources, and these marginal costs must rise at the appropriate rate over time. Allocating emissions allowances in appropriate quantities creates a market and market price that performs the same signalling function as a tax as long as all carbon is covered in all regions and human activities. Generating the appropriate rate of price escalation could be achieved by managing the supply of allowable permits. Regulatory policy instruments could also be employed to achieve the same end. However, it can be difficult in practice to maintain equality in the marginal value of GHG emissions across sectors employing regulatory instruments alone. Each policy instrument has its own set of advantages and disadvantages and the eventual choice will depend on many factors, including the local institutional history and context, and will probably evolve with time (DOE, 1989).

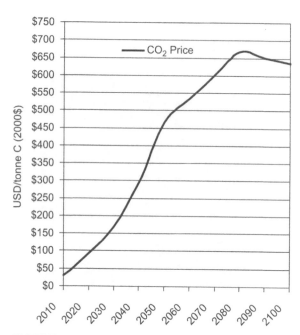

FIGURE 6 Price path for carbon emissions consistent with stabilization of radiative forcing at 3.4 W/m².

Emissions mitigation relative to the reference scenario and relative to the present is substantial in both the USA and globally (Figure 7). By 2050, global emissions are 50% of 2010 levels. By 2100, global emissions have declined by two-thirds relative to 2010 and by almost 90% relative to the reference scenario. The USA's emissions mitigation is somewhat greater relative to 2010 than the global average, with a 58% reduction in 2050 rather than 50%. Because the rate of emissions growth in the USA is slower than for the world as a whole, its emissions mitigation relative to the reference scenario is somewhat smaller than the world average. The degree of emissions mitigation and its timing are, nonetheless, daunting.

5. The US energy system and stabilization at 3.4 W/m²

5.1. Overview of the US energy system

Like the rest of the world, the US energy system is dominated by fossil fuel use at present, and in the reference scenario that dominance persists throughout the century (Figure 5) despite a growing share of energy provided by non-emitting energy sources. In contrast, dramatic changes occur in both the US and global energy systems in the 3.4 W/m² stabilization case. By 2050, fossil fuel use has declined to about half of the USA's primary energy consumption (Figure 8A). In 2050, the USA's primary energy consumption is about 20% smaller than in the reference scenario. These reductions in demand are driven by higher energy prices engendered by the price on GHGs.

By mid-century almost half of all primary energy is provided by non-fossil energy forms, largely nuclear, solar, wind and biomass. Biomass energy is treated as a non-carbon-emitting energy form. The increased use of nuclear energy is particularly striking. Power production from nuclear energy doubles by 2040 and triples by 2070, but its relative contribution remains stable in the final decades of the century. The increase in deployment of non-biomass renewable energy

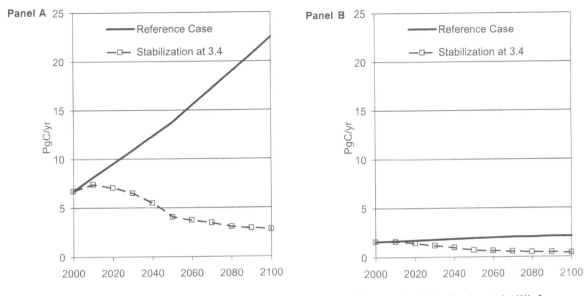

FIGURE 7 Global (A) and US carbon emissions (B) consistent with stabilization of radiative forcing at 3.4 W/m².

production is even more striking. Production more than triples by 2040 and quadruples by 2070, though much of the increase in market share also occurs in the reference scenario. Changes in primary energy consumption are shown in Figure 8B plus fossil fuel energy used in conjunction

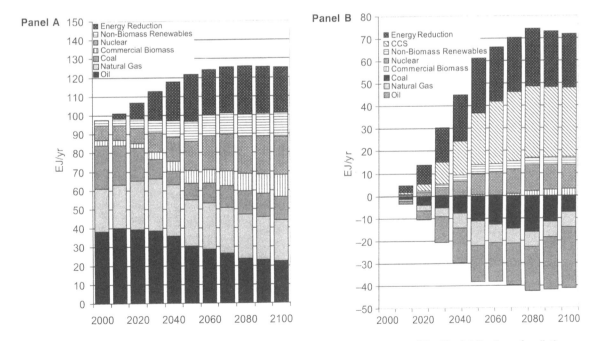

FIGURE 8 The USA's primary energy consumption (A) and associated changes (B) with stabilization of radiative forcing at 3.4 W/m².

with CO_2 capture and geological storage (CCS), and reductions in aggregate energy use. Negative values refer to reductions in energy use by fuel, while positive values are increased primary energy consumption in the stabilization case relative to the reference scenario. (Note that overall reductions in primary energy consumption are shown as a positive value.)

5.2. CO_2 capture and storage

An important technology option that is assumed to be available in the scenario is CO_2 capture and storage technology. Direct use of fossil fuels declines steadily in the stabilization scenario. The use of coal declines precipitously and remaining coal use is increasingly deployed with CO_2 capture and geological storage (CCS) technology. By 2050 all coal use employs CCS, and by 2070 almost all fossil fuel power generation employs CCS technology (see Figure 9).

It is assumed that CO_2 can be scrubbed, transported and permanently stored in geological repositories. Given that all fossil power generation is assumed to be able to employ this technology it is important to acknowledge that many important steps remain to be taken before this can be accomplished. While there is good experience with all of the components of CCS technology, the technology has not been deployed on a large scale. Projects in operation today capture and store approximately 1 TgC/yr (3.67 $TgCO_2$) globally (Dooley et al., 2006). However, in the stabilization scenario 55 TgC/yr are stored by 2020 in the USA alone (Figure 10A) while 267 TgC/yr are stored globally. By the middle of the century, deployment has increased by another order of magnitude (Figure 10B), engendering a further set of challenges in terms of the scale of the necessary technology deployment.

Since CCS is presently not deployed on a large scale, a rapid ramp-up implies addressing important transition issues such as the availability of drilling rigs, site characterizations, manufacture of capture

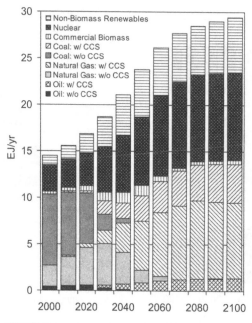

FIGURE 9 The USA's power generation by technology with stabilization of radiative forcing at 3.4 W/m².

and transport systems, and establishment of the institutional mechanisms necessary to facilitate and regulate deployment. At the most fundamental level, CCS technology creates additional costs for any productive activity. Thus, unless carbon takes on a sufficiently high value, either explicitly or implicitly, the technology will not be deployed. In addition, the technology is associated with long-lived capital assets, and therefore it must be possible for investors to form expectations about future values of carbon that are consistent with the path outlined earlier in this article.

Instruments to recognize the emissions reductions that occur if CO_2 is captured and stored need to be developed. We have also assumed that captured carbon that is stored in a geological reservoir remains there indefinitely. Monitoring and verification will be an essential component of successful technology deployment. Finally, no technology can ever be perfect. Thus, instruments will need to be created that allow investors to manage their long-term risks.

Globally, the challenge of technology development and deployment and the development of associated monitoring mechanisms loom large. Cumulative carbon capture reaches 32 PgC by 2095 in the USA and almost 250 PgC globally. At a coarse scale, this is well within the magnitude of maximum geological storage potential estimated for the USA (>1,000 PgC) and for the world (>2,800 PgC) (Edmonds et al., 2007). However, some regions, such as Japan and Korea, may find limited geological storage potential within their national boundaries. Even within the USA, storage potential is not evenly distributed. Power generators in the Ohio valley will find relatively abundant opportunities. However, power generators in New England may find fewer potentially attractive sites (Dooley et al., 2006).

CCS technology is potentially applicable to many large point-source emitters. The largest application of the technology in the analysis reported here was in electric power generation. However, other point-source emissions are also important, including cement kilns, iron and steel foundries, natural gas processing, petroleum refineries and, potentially, hydrogen production facilities.

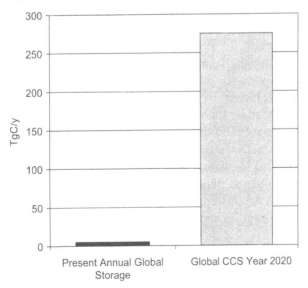

Panel A. Present and year 2020 global deployment rates

FIGURE 10 Deployment of CCS with stabilization of radiative forcing at 3.4 W/m².

Panel B. U.S. Deployment between 2020 and 2095.

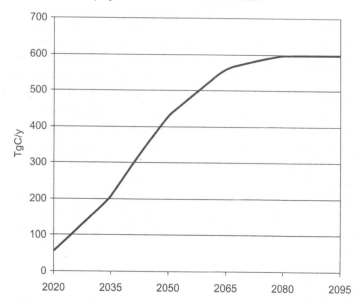

FIGURE 10 The USA's deployment of CCS with stabilization of radiative forcing at 3.4 W/m². (*Cont'd*)

5.3. Bioenergy and terrestrial carbon

Globally there is approximately 1,800 PgC stored in terrestrial ecosystems. For comparison, cumulative carbon emissions in the reference scenario in the 21st century were approximately 1,400 PgC. On balance, terrestrial reservoirs are believed to be accumulating carbon through regrowth and CO_2 fertilization effects (see Figure 11). The net uptake by ecosystems is thought to be more than sufficient to offset present carbon emissions from land-use change, which amount to approximately 1.5 PgC/year.

Figure 11 shows three time-paths for global net terrestrial carbon system uptake (shown as negative emissions in Figure 11), one time-path for the reference scenario, and two alternative time-paths for net terrestrial carbon system uptake for stabilization of radiative forcing at 3.4 W/m². Carbon uptake along the reference time-path is higher for two reasons; the first of which is the CO_2 fertilization effect whereby plant growth is enhanced by higher atmospheric CO_2 levels. The CO_2 concentration is higher in any year along the reference scenario time-path, so plants store more carbon in biomass and soils. Second, the extent of commercial biomass cropping is smaller in the reference scenario, which in turn means that the demand for land is smaller than in the stabilization scenario. The larger demand for land for biomass crops in the policy scenario results in additional carbon releases through deforestation.

Note that these two effects are offset by temperature feedbacks, which are thought to result in a net reduction in carbon storage as temperatures increase. Temperature feedbacks are larger in the reference case; an effect that acts in the opposite direction to CO_2 fertilization and enhanced deforestation, due to increased biomass demand.

The important distinction between the two scenarios that stabilize radiative forcing at 3.4 W/m² is that along the path with higher terrestrial carbon cycle uptake, the line labelled '3.4 W/m²' in Figure 11, a value is placed on terrestrial carbon, while along the line labelled '3.4 W/m² No Carbon Tax on Land-use Change Emissions' in Figure 11 the value of terrestrial carbon is zero.

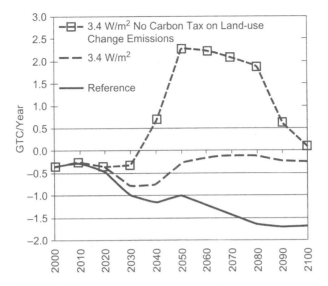

FIGURE 11 Impact on global net terrestrial carbon uptake of a tax on terrestrial carbon emissions.

The reason that the terrestrial reservoirs shift from sink to source between the line labelled '3.4 W/m²' and the line labelled '3.4 W/m² No Carbon Tax on Land-use Change Emissions' can be traced to bioenergy. Bioenergy is treated as non-carbon emitting, because it obtains its carbon from the atmosphere. However, bioenergy derived from purpose-grown crops requires land. As the price of carbon rises, the relative attractiveness of bioenergy relative to fossil fuels grows. If terrestrial carbon is not valued, then excessive land is allocated to the production of bioenergy and insufficient land is allocated to storage of carbon in forests and soils. Figure 11 shows that the failure to value terrestrial carbon can lead to dramatic rates of deforestation to obtain land, especially in the tropics, for bioenergy plantations.

Because the atmosphere treats all carbon equally, the value associated with emissions of carbon to the atmosphere should be the same regardless of the source. Thus, despite the difficulties associated with placing a value on terrestrial carbon, valuing that carbon at zero could potentially lead to substantial negative impacts of climate policies, including increased deforestation, decreased afforestation and reforestation, and decreased provision of ecosystem services.

5.4. End-use energy and electrification

Electricity plays an important role in the stabilization scenario. Even in the reference scenario the ratio of electricity to total end-use energy consumption rises over time, just as it rose historically. However, the ratio rises significantly more rapidly in the stabilization scenario (Figure 12).

As the price of carbon rises, the power sector relies increasingly on non-emitting technologies (Figure 9). End-use sectors, particularly buildings and industry, shift increasingly to electricity and away from the direct use of fossil fuels as the carbon price rises. For example, building sector emissions are driven down by more than 65% in the 3.4 W/m² stabilization scenario relative to the reference scenario in the year 2095. Yet, energy use declines only 20%. Forty-five percent of emissions reductions in the buildings sector are the consequence of fuel switching, primarily to electricity. By 2095, electricity is responsible for 80% of energy consumed in the buildings sector in 2095.

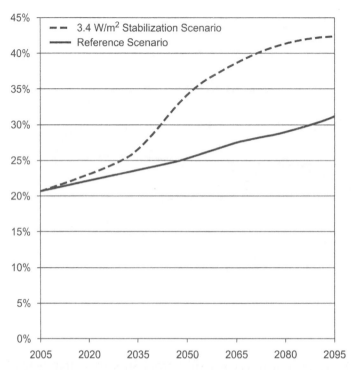

FIGURE 12 Ratio of electricity consumption to total final energy consumption reference and stabilization at radiative forcing at 3.4 W/m²

The role of technology in end-use sectors shifts over time in the 3.4 W/m² stabilization scenario. Initially technologies that conserve electricity reduce emissions dramatically by reducing power production that would have been predominantly generated with fossil fuels. As the power sector decarbonizes, technologies that can substitute electricity for fossil fuels in the provision of energy services become increasingly valuable.

Edmonds et al. (2005) showed that this phenomenon has important policy implications. Policies that value carbon in power generation, but which do not value carbon in end-use sectors, rapidly lead to economic inefficiencies that raise the cost of stabilization. They do so because valuing carbon in power generation raises the price of electric power. If carbon is not valued in end-use sectors, then the price of electricity rises relative to fossil fuels and end-use sectors substitute fossil fuels for electricity and de-electrify. As a consequence, the decarbonizing power sector is employed relatively less in end-use, and carbon-emitting technologies are employed relatively more. This is yet another example of the principle of valuing all carbon equally at the margin throughout the economy, regardless of sector.

5.5. Technology and cost in the near, mid- and long term

The present price of carbon and other GHGs depends on the entire period. In the analysis reported here the relationship is simple and direct. We assumed a price trajectory that minimizes social cost. The price of carbon and all GHGs rises at the rate of interest plus the rate of removal from the stock in the atmosphere. Thus, every future price is determined by the choice of the present price.

Of course, the real world does not work in an ideal manner, nor is the future perfectly predictable. There is no way of knowing either the future or the suite of technologies that will be available in the future. The choice of the present price of carbon depends on expectations about the future which are subject to regular reassessment and revision as the future unfolds.

That having been said, the choice of the initial price depends on expectations about future technology. A more pessimistic outlook for future technology performance and availability implies a higher current price of carbon (and other GHGs). Similarly, an optimistic view of future technology performance and availability implies a lower current price of carbon.

Any mitigation programme begins in the present, deploying the technology that is available. As the stabilization regime moves forward in time, the opportunity exists to improve the present suite of technologies. Investments in present technology and in the creation of incremental improvements in that suite of technologies will be driven not only by the present carbon price, but the expectation that the price will rise with time. Both the existence of a value on carbon and other GHGs and the creation of an expectation for future price increases consistent with stabilization are important elements in stabilizing radiative forcing at 3.4 W/m².

As important as near-term prices and expectations are in stabilizing radiative forcing at 3.4 W/m², it should be noted that the bulk of emissions mitigation occurs not in the near or mid-term, but in the post-2050 period. Thus, another important element in a technology strategy needs to be investment in basic science that holds the potential to provide the foundation upon which future technologies can be built. Managing technology risk in a world where 'spillover' effects are pervasive in technology development (Clarke and Weyant, 2002; Clarke et al., 2006) implies investments in a broad spectrum of scientific inquiry.

The price of CO_2 rises to steadily reduce global carbon emissions (see Figure 13). The rise is exponential until 2050, at which point the concentration of CO_2 has reached 450 ppm and radiative forcing has reached its steady-state level. Thereafter the price of carbon is determined by the commitment to maintain radiative forcing at 3.4 W/m². The price of carbon continues to rise until 2080, after which point it begins to decline as technology advance and the declining rate of physical carbon emission reduction requirements finally reduce the marginal cost. Total present discounted costs for the USA to limit carbon emissions amount to approximately 1.2 trillion year 2000 constant US$, compared with

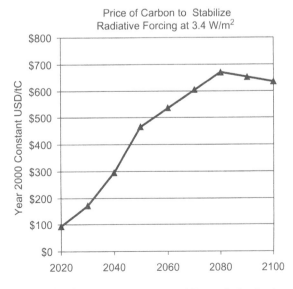

FIGURE 13 Price of carbon to stabilize radiative forcing at 3.4 W/m²

global total present discounted costs of 9.8 trillion year 2000 constant US$, where costs are discounted at 5% per year. These costs are the lowest potential costs for achieving the long-term stabilization goal and reflect the idealized representation of the policy, namely a world in which carbon emissions are reduced in a globally and intertemporally efficient emissions limitation regime. Relaxation of this assumption raises costs both globally and in the USA, potentially by several orders of magnitude.

6. Delays in participation

The scenario we have analysed, which stabilizes radiative forcing at 3.4 W/m^2, was constructed on the assumption that the world undertakes actions in an economically efficient manner. All regions of the world were assumed to participate in the global emissions control regime beginning in the year 2012, and applied the same price to carbon and other GHGs, with that price rising at the economically efficient rate. Relaxing these assumptions has substantial implications for the USA if the goal of stabilizing radiative forcing at 3.4 W/m^2 is still to be achieved.

As discussed earlier, limiting climate change is inherently a global, not regional, problem. The concentration of GHGs in the atmosphere depends on all GHG emissions everywhere and from all sources. Stabilizing the concentration of CO_2 in the atmosphere also entails limiting cumulative, not annual, carbon emissions everywhere and from all sources. This means that limiting radiative forcing to 3.4 W/m^2 implies a limit in cumulative emissions. Delayed participation on the part of any region means that the emissions mitigation that occurs in the idealized control regime analysed in this article must be made up by other regions.

As shown by Edmonds et al. (2007), there is little room for intertemporal displacement of emissions into the future for a 3.4 W/m^2 target, meaning that delays in the participation of some regions mean substantial increases in near-term emissions reductions for those countries acting first. For example, in Edmonds et al. (2007), if the non-Annex I countries were to delay emissions reductions to 2020 or beyond, the USA might need to reduce emissions by as much as 50% by 2020 to help keep the globe on an emissions pathway that would allow the 3.4 W/m^2 limit to be met. Reductions of such magnitude, while technically not impossible, are so drastic that they would stress the ability of US society to actually accomplish such reductions.

The point is also general and symmetrical. That is, delays on the part of the USA would shift the burden to participating regions, increasing the cost that mitigating regions would experience if the 3.4 W/m^2 goal were to be realized.

Climate change is a public-good problem, and thus there is always an incentive for any party to under-report their desire to reduce emissions and to be a 'free rider'. The 3.4 W/m^2 limit is so severe that there is little latitude to shift emissions mitigation into the future to compensate for delayed participants.

Limiting the change in radiative forcing to 3.4 W/m^2 is an enormous and unprecedented global and regional challenge. Delayed participation on the part of any major region shifts the burden of emissions mitigation onto regions that are mitigating emissions. And, if the non-participating regions account for a significant share of emissions, a point is rapidly reached beyond which it is physically impossible to limit radiative forcing to 3.4 W/m^2 without first 'overshooting' the limit for some period of time.

7. Conclusions

Stabilization presents a daunting challenge for all countries of the world, regardless of their stage of development, institutions or technological capabilities. This article has explored the implications for the USA of stabilization at 3.4 W/m^2. Stabilization at this level, even under idealized conditions of nearly immediate global cooperation, will require a transformation of the US energy system beginning almost immediately and extending throughout the century and beyond. This transformation will need to be even more rapid and extensive if the emissions reduction regime encompasses only a portion of the global economy.

The availability of advanced technologies such as CCS, sustainable bioenergy production, wind and solar, nuclear energy and end-use efficiency improvements will facilitate this transition. Indeed, the degree to which technology advances over the coming century is among the most important determinants of the economic costs of stabilization for the USA and the rest of the world. The scope of the energy system transformation highlights the need to begin deploying technologies that are currently available and to continue to invest in R&D to develop newer, more efficient, and less expensive low- or zero-carbon energy supply technologies and end-use technologies.

The rapid pace of the transition necessary to meet a 3.4 W/m² target, particularly in what may be less idealized cases, where coordinated, global action does not immediately occur, raises questions about the ability of even wealthy societies to achieve the requisite widespread changes over a period of just one or two decades. The scale of expansion of CCS technology seen in this scenario is only one example of the technological and institutional changes that will be needed to effect such a transformation. While model results such as this can illustrate a potential path of technology deployment that would lead to a given goal, the process of initial technology development and deployment, in particular, is not explicitly modelled. It is not known, for example, if the roughly 100-fold expansion in global CCS activities projected in this scenario to occur over the next 12 years is even possible. If such a rapid deployment of this particular technology is not possible, then mitigation activities in other sectors would need to be further accelerated to achieve the same goal; other sectors that have their own institutional and logistical barriers. Further work is needed in order to better understand and quantify both the constraints on expansion of mitigation technologies for particular sectors and the methods that could be used to reduce these constraints.

Acknowledgements

The authors are grateful to the US Department of Energy's Office of Science and to the Electric Power Research Institute for financial support for the research whose results are reported here. The authors are also grateful to the anonymous *Climate Policy* referees for helpful comments on an earlier draft of this article. The views expressed in this article are those of the authors alone and do not represent those of any other institution. Any errors that remain are the authors' alone.

Notes

1. Radiative forcing could also be changed through direct anthropogenic intervention with incoming solar radiation through geoengineering, although this is not considered here. Changes in solar output would also alter forcing, although the estimated historical changes in solar forcing are much smaller than anthropogenic forcings to date (Forster et al., 2007).
2. For additional information see www.globalchange.umd.edu/models/MiniCAM/.
3. This is a standard methodological approach that has been employed by such studies as the IPCC *Special Report on Emissions Scenarios* (Nakicenovic and Swart, 2000), and the earlier IPCC IS92 scenarios (Leggett et al., 1992).
4. A variety of alternative scenarios have been developed including those with both higher and lower radiative forcing. See Nakicenovic and Swart (2000) or the more recent Van Vuuren et al. (2007).
5. The underlying technical assumptions that lead to this conclusion are documented in Clarke et al. (2007b).
6. Note that until the concentration reaches its steady-state level the economically efficient price rises at the rate of interest plus the average rate of ocean uptake (Edmonds et al., 2008). Once the concentration of CO_2 reaches its steady-state level, emissions rates, including both net terrestrial and industrial, are determined solely by rate of ocean uptake. Prices need no longer rise exponentially.

References

Brenkert, A., Smith, S., Kim, S., Pitcher, H., 2003, *Model Documentation for the MiniCAM*, PNNL-14337, Pacific Northwest National Laboratory, Richland, WA.

Clarke, L., Weyant, J., 2002, 'Modeling-induced technological change: an overview', in: A. Grubler, N. Nakicenovic, W.D. Nordhaus (eds), *Technological Change and the Environment*, Resources for the Future, Washington, DC.

Clarke, L., Weyant, J., Edmonds, J., 2006, 'The sources of technological advance: what do the models assume?' *Energy Economics* 28(5–6), 579–595.

Clarke, L., Edmonds, J., Jacoby, H., Pitcher, H., Reilly, J., Richels, R., 2007a, *CCSP Synthesis and Assessment Product 2.1, Part A: Scenarios of Greenhouse Gas Emissions and Atmospheric Concentrations*, U.S. Government Printing Office, Washington, DC.

Clarke, L., J. Lurz, M. Wise, J. Edmonds, S. Kim, S. Smith, H. Pitcher, 2007b, *Model Documentation for the MiniCAM Climate Change Science Program Stabilization Scenarios: CCSP Product 2.1a*, PNNL Technical Report, PNNL-16735.

DOE (U.S. Department of Energy), 1989, *A Compendium of Options for Government Policy to Encourage Private Sector Responses to Potential Climate Change*, DOE/EH-0102 and DOE/EH-0103, National Technical Information Service, U.S. Department of Commerce, Springfield, VA.

Dooley, J.J., Dahowski, R.T., Davidson, C.L., Wise, M.A., Gupta, N., Kim, S.H., Malone, E.L., 2006, *Carbon Dioxide Capture and Geologic Storage: A Core Element of a Global Energy Technology Strategy to Address Climate Change*, Joint Global Change Research Institute (JGCRI), Battelle Memorial Institute, College Park, MD.

Edmonds, J., Reilly, J., 1985, *Global Energy: Assessing the Future*, Oxford University Press, New York.

Edmonds, J., Wilson, T., Wise, M., Weyant, J., 2005, 'Electrification of the economy and CO_2 emissions mitigation,' *Journal of Environmental Economics and Policy Studies* 7, 175–203.

Edmonds, J., Dooley, J., Kim, S., Friedman, S., Wise, M., 2007, 'Technology in an integrated assessment model: the potential regional deployment of carbon capture and storage in the context of global CO_2 stabilization', in: M. Schlesinger, F.C. de la Chesnaye, H. Kheshgi, C.D. Kolstad, J. Reilly, J.B. Smith, T. Wilson (eds), *Human-Induced Climate Change: An Interdisciplinary Perspective*, Cambridge University Press, Cambridge, UK.

Edmonds, J., Clarke, L., Lurz, J., Wise, M., 2008, 'Stabilizing CO_2 concentrations with incomplete international cooperation', *Climate Policy* 8 (forthcoming).

Forster, P., Ramaswamy, V., Artaxo, P., Berntsen, T., Betts, R., Fahey, D.W., Haywood, J., Lean, J., Lowe, D.C., Myhre, G., Nganga, J., Prinn, R., Raga, G., Schulz, M., Van Dorland, R., 2007, 'Changes in Atmospheric Constituents and in Radiative Forcing', in: S. Solomon, D. Qin, M. Manning, Z. Chen, M. Marquis, K.B. Averyt, M. Tignor, H.L. Miller (eds), *Climate Change 2007: The Physical Science Basis*. Contribution of Working Group I to the Fourth Assessment Report of the Intergovernmental Panel on Climate Change, Cambridge University Press, Cambridge, UK.

Gillingham, K.T., Smith, S.J., Sands, R.D., 2007, 'Impact of bioenergy crops in a carbon dioxide constrained world: an application of the MiniCAM energy-agriculture and land use model', *Mitigation and Adaptation Strategies for Global Change* doi: 10.1007/s11027-007-9122-5.

Hotelling, H., 1931, 'The economics of exhaustible resources', *Journal of Political Economy* 39, 137–175.

Kim, S.H., Edmonds, J., Lurz, J., Smith, S.J., Wise M., 2006, 'The ObjECTS Framework for Integrated Assessment: hybrid modeling of transportation', *Energy Journal* (Special Issue 2), 51–80.

Leggett, J., Pepper, W.J., Swart, R.J., Edmonds, J., Meira Filho, L.G., Mintzer, I., Wang, M.X., Wasson, J., 1992, 'Emissions scenarios for the IPCC: an update', in: *Climate Change 1992: Supplementary Report to the IPCC Scientific Assessment*, Cambridge University Press, Cambridge, UK.

Nakicenovic, N., Swart, R. (eds), 2000, *Special Report on Emissions Scenarios*, Cambridge University Press, Cambridge, UK.

Peck, S.C., Wan, Y.S., 1996, 'Analytic solutions of simple greenhouse gas emission models', in: E.C. Van Ierland, K. Gorka (eds), *Economics of Atmospheric Pollution*, Springer, New York.

Raper, S.C.B., Wigley, T.M.L., Warrick, R.A., 1996, 'Global sea-level rise: past and future', in: J.D. Milliman, B.U. Haq (eds), *Sea-Level Rise and Coastal Subsidence: Causes, Consequences, and Strategies*, Kluwer Academic Publishers, Dordrecht, The Netherlands, 11–45.

Van Vuuren, D.P., Meinshausen, M., Plattner, G.-K., Joos, F., Strassmann, K.M., Smith, S.J., Wigley, T.M.L., Raper, S.C.B., Riahi, K., de la Chesnaye, F., den Elzen, M., Fujino, J., Jiang, K., Nakicenovic, N., Paltsev, S., Reilly, J.M., 2007, *Stabilization of 21st Century Climate*, Working Paper, MNP, The Netherlands.

Wigley, T.M.L., Raper, S.C.B., 1992, 'Implications for climate and sea-level of revised IPCC emissions scenarios', *Nature* 357, 293–300.

Wigley, T.M.L., Raper, S.C.B., 2002, 'Reasons for larger warming projections in the IPCC Third Assessment Report', *Journal of Climate* 15, 2945–2952.

Permit sellers, permit buyers: China and Canada's roles in a global low-carbon society

CHRIS BATAILLE*, JIANJUN TU, MARK JACCARD

Energy and Materials Group, School of Resource and Environmental Management, Simon Fraser University, 8888 University Drive, Burnaby, BC, Canada V5A 1S6

The challenge of creating a global low-carbon society is examined from the perspectives of a slow-growing but highly developed economy (Canada) and a fast-growing developing economy (China). Both countries' responses are compared to a similar carbon price schedule (US$10/tCO$_2$e in 2013 rising exponentially to $100 by 2050) using a hybrid technologically explicit and behaviourally realistic model with macroeconomic feedbacks (CIMS). Then additional measures are imposed based on the national circumstances of each country; for Canada we simulate a 50% reduction by 2050, and stabilization for China. The scale of the challenge in all cases requires that every available option be vigorously pursued, including energy efficiency, fuel switching, carbon capture and storage, and accelerated development of renewables; to compensate, there are significant co-reductions of local air pollutants such as SO$_x$ and NO$_x$. Finally, the abatement cost schedules of China and Canada are compared, and implications considered for carbon permit flows if the cost schedule of the rest of the developed world is assumed to be similar to that of Canada. We found that the developed world and China could collectively reduce emissions by 50% in 2050 at a price of $175/tCO$_2$e, with permits flowing from the developed countries to China; while abatement costs are lower in China up to $75/t, at higher prices reductions are less costly in the developed world. Our results indicate that a global low-carbon society is feasible, on condition that policy makers are willing and able to impose long-term, credible policy packages with carbon pricing policy as the core element, coupled with supplementary regulations to address market failures.

Keywords: Canada; carbon trading; China; development; endogenous technical change; hybrid model; low-carbon society; public policy; scenario modelling; stabilization pathways

Le défi pour la création d'une société mondiale sobre en carbone est examiné selon la perspective d'une économie à faible croissance mais fortement développée (le Canada) et une économie en développement à forte croissance (la Chine). La réponse de ces deux pays à un même barème des prix du carbone est comparée (US$10/tCO$_2$eq en 2013 en croissance exponentielle jusqu'à $100 d'ici 2050), sur la base d'un modèle hybride technologiquement précis et réaliste en comportements avec feedbacks macroéconomiques (modèle CIMS). Des mesures supplémentaires sont ensuite imposées sur la base de circonstances nationales propres à chaque pays; pour le Canada nous simulons une réduction de 50% d'ici 2050, et une stabilisation pour la Chine. L'ampleur du défi dans tous les cas de figure nécessite une poursuite vigoureuse de chaque option disponible, y compris de l'efficacité énergétique, du changement de combustibles, de la capture et stockage du carbone, et du développement accéléré des énergies renouvelables; pour compenser, de considérables co-réductions de pollutions de l'air local sont obtenues, tels qu'en SO$_x$ et NO$_x$. Finalement, les barèmes des coûts de réduction pour la Chine et le Canada sont comparés, et leur incidence sur les flux de permis de carbone est considérée, le barème des prix pour le reste du monde développé étant supposé du même ordre que pour le Canada. Nous montrons qu'à un prix de $175/tCO$_2$eq, le monde développé et la Chine pourraient ensemble réduire les émissions de 50% d'ici 2050, les permis allant des pays développés vers la Chine; alors qu'en Chine les couts de réduction sont plus faibles (jusqu'à $75/t), dans le monde développé les réductions sont moins chères pour des prix plus élevés. Nos résultats indiquent qu'une société mondiale sobre en carbone est possible, à condition que les décideurs soient disposés et capables d'imposer des paquets de politiques à long terme crédibles,

■ *Corresponding author. E-mail*: cbataill@sfu.ca

CLIMATE POLICY 8 (2008) S93–S107

doi:10.3763/cpol.2007.0494 © 2008 Earthscan ISSN: 1469-3062 (print), 1752-7457 (online) www.climatepolicy.com

ayant comme élément principal la fixation du prix du carbone, associé à des réglementations supplémentaires pour résoudre les failles de marché.

Mots clés: axes de stabilisation; Canada; changements technologiques endogènes; Chine; développement; échange de quotas de carbone; modèle hybride; modélisation de scénarios; politiques publiques; société sobre en carbone

1. Introduction

An international consensus is forming around the need to reduce global greenhouse gas emissions (GHGs) by a substantial amount in order to avoid the risk of disruptive climate change. This is anticipated to occur should the atmospheric concentration of GHGs, which is currently 380 parts per million (ppm) and 100 ppm above its pre-industrial level, exceed 450–650 ppm.[1] Experts are debating the necessary emissions reductions, but calls for a 50% drop from today's emissions levels by 2050 are common, implying the creation of a 'low-carbon society' (LCS).[2] But how do we achieve such a fundamental change in the nature of our economies without unpalatable sacrifices with regard to our standard of living? There is general agreement among economists that application of a sufficient price signal on GHGs, either in the form of a GHG tax or an emissions cap with tradable permits, is the most effective and efficient way to reduce emissions, as it allows emitters to decide who will act and how to respond. There is also a growing consensus that supplementary policies (e.g. in the form of command and control regulations such as efficiency standards for buildings and vehicles) will be required to cover sectors not adequately influenced by the price signal, or otherwise subject to market failures (Stavins, 1998; Stavins et al., 2007).

In this article we look at the LCS challenge from the perspectives of a slow-growing but highly developed economy (Canada) and a fast-growing developing economy (China). We first compare both countries' responses to a similar GHG price schedule (US$10/tCO$_2$e in 2013 rising exponentially to $100 by 2050) using a hybrid technologically explicit and behaviourally realistic model with macroeconomic feedbacks (CIMS). We then impose additional measures for both countries; these measures are based on the national circumstances of each, and thus are distinctly different. Finally, we discuss the implications of these results, including those for international development and potential emissions credit trading flows.

2. The CIMS hybrid technology simulation model

The model we use for this analysis, CIMS, is an integrated, energy–economy equilibrium model that simulates the interactions between energy supply, demand and the macroeconomic performance of any economy, including trade effects. It is designed to model energy and emissions policies, including emissions taxes, cap-and-trade systems, technology regulations and subsidies. It specializes in simulating policies that include both broad pricing instruments and technology-specific regulations, a combination difficult to model by both standard bottom-up technology models and top-down computable general equilibrium (CGE) or macro-econometric models.

CIMS is of a class of hybrid models that bring together technological detail (historically an exclusive characteristic of bottom-up models), realistic firm and consumer investment and consumption behaviour, and macroeconomic feedbacks (historically characteristics of top-down macro-econometric or CGE models) (Hourcade et al., 2006). CIMS is a bottom-up model that approaches hybridization by the use of empirically estimated parameters to model micro-level firm and consumer investment consumption behaviour, as well as basic macroeconomic feedbacks that operate in response to changes in price caused by changes in the cost of producing energy and emissions-intense goods and services. Other models, starting from the top-down CGE or macro-econometric perspective, instead add technological detail via detailed energy supply models, the use of highly disaggregated and technology-specific production functions, specialized mathematical techniques to simulate discrete technologies, and functions to include endogenous technical change via learning-by-doing and R&D. For an introduction to the evolving art of hybrid modelling, including several examples, refer to the special issue of the *Energy Journal* (Hourcade et al., 2006).

Starting with an initial exogenous forecast of physical output, CIMS sector sub-models (Table 1) track the evolution of all energy-using capital stocks over a 50-year time horizon in a baseline and under the influence of policy, including base stock, new purchases, retrofits, and retirement at the individual technology level. When making new purchases and performing retrofits, consumers and firms make decisions with limited foresight based on financial and intangible costs. These include: financial costs such as capital, labour, energy, material and emissions charges; intangible costs such as risk, time preference, option value, and individual and firm level positive and negative externalities. The model also includes two functions for simulating endogenous change in individual technologies' characteristics in response to policy: a declining capital cost function to represent economies of scale and learning-by-doing, and a declining intangible cost function for new technologies that may be unfamiliar to firms and consumers.

Approximately 2,800 technologies are competing in CIMS to meet the demand for hundreds of final and intermediate goods and services. These competitions are organized by sector; the most important final and intermediate goods and services are described in Table 1, but there are hundreds of additional intermediate end-uses in CIMS, e.g. space heating and cooling, pumping, compression, conveyance, steam, air displacement, etc.

A simulation iterates between the energy supply and demand sectors until energy price variation falls below a threshold percentage, and repeats this convergence procedure to find the supply/demand equilibrium in each subsequent 5-year period of a complete run. A similar iterative convergence procedure is followed to equilibrate the markets for goods and services. While CIMS does not explicitly include a government sector that taxes and produces public goods in return, output-weighted revenue recycling is an option for emissions pricing policies. The version of CIMS used for this project does not equilibrate government budgets, nor markets for employment and investment. A CGE partner model for CIMS, which will include these dynamics, is in development. See Bataille et al. (2006) for an extended description of CIMS.

All simulations include baseline and policy accounting of capital stock, energy use and emissions. CIMS tracks production, transmission, distribution and end-use of all energy types, including electricity, ethanol, hydrogen, different refined petroleum products, several types of coal, natural gas and renewables. In terms of emissions, CIMS tracks all combustion and process CO_2, process and fugitive CH_4, process SF_6, and most sulphur oxides (SO_x), nitrogen oxides (NO_x), volatile organic compounds (VOCs), carbon monoxide (CO) and particulate matter (TPM, $PM_{2.5}$ and PM_{10}). Because of its focus on energy, climate change and air quality policy, there is a special emphasis on fuel switching, energy-efficiency investment opportunities, renewable energy, biofuel and hydrogen supply/demand infrastructure and technologies, carbon capture and storage (CCS), and nuclear energy. International aviation and marine emissions are not included, given the focus of the current CIMS models on nation states, but CIMS-Global, which is under development, may include this as option.

TABLE 1 CIMS sub-sectors

Sector models	End-uses or products of the sector models[a]
Commercial/institutional	Space heating/cooling, refrigeration, cooking, hot water, plug load
Transportation	Freight (marine, road and rail); Personal (intercity/urban, split into single- and high-occupancy vehicles, public transit/walking/cycling); Off-road
Residential	Space heating/cooling, refrigeration, dishwashers, freezers, ranges, clothes washers and dryers, other
Iron and steel	Slabs, blooms and billets
Pulp and paper	Newsprint, linerboard, uncoated and coated paper, tissue, market pulp
Metal smelting	Lead, copper, nickel, titanium, magnesium, zinc and aluminium
Chemical production	Chlor-alkali, sodium chlorate, hydrogen peroxide, ammonia, methanol, polymers
Mining	Open-pit, underground, potash
Industrial minerals	Cement, lime, glass, bricks
Other manufacturing	Edible goods, rubber, plastics, leather, textiles, clothing, wood products, furniture, printing, machinery, transportation equipment, electrical and electronic equipment
Petroleum refining	Gasoline, diesel, kerosene, naptha, aviation fuel, petroleum coke
Electricity production	Electricity
Natural gas production	Natural gas and natural gas liquids
Coal mining	Lignite, sub-bituminous, bituminous and anthracite coal
Crude oil production	Light/medium and heavy crude oil, bitumen and synthetic crude oil

[a] Includes space heating and cooling, pumping, compression, conveyance, hot water, steam, air displacement, and motor drive as applicable.

There are several regional and national versions of the CIMS model: CIMS-Canada, CIMS-China, CIMS-North America, and a global CIMS under construction. The two models used for this project, CIMS-Canada and CIMS-China, are the most developed; CIMS-Canada is used regularly by Canadian policy makers for analysis of potential energy and GHG policies, and is therefore the most disaggregated (7 regions) and detailed. All the CIMS models disaggregate into at least 15 separate sub-sector models (Table 1). CIMS requires an initial physical baseline output forecast by sector for each sub-region, and inputs of baseline energy prices; for Canada this is Natural Resources Canada's *Canada's Energy Outlook 2006* (Natural Resources Canada, 2006), and for China this is an update of Tu et al. (2007). Energy prices can be adjusted endogenously to reflect policy to the extent that the market price for various energy commodities is set internally to the geographical scope of the model (e.g. electricity prices in Canada are generally set regionally, natural gas prices are set within a North American market with a backstop price for LNG, and crude oil prices at the global clearing price).

3. Simulation method

We focus on three different carbon futures in this study: a *Base case* where no carbon restrictions are imposed; a *Carbon price* scenario where there is a global price of US $10/tCO$_2$e in 2013 rising to

$100/tCO$_2$e in 2050; and a *Carbon-plus* scenario where additional policies coupled with carbon price are implemented to achieve a LCS goal.

3.1. Base case: No global carbon constraint is imposed

Our Base case is based on Canada's official national energy use forecast[3] and updated reference energy projection in Tu et al. (2007).[4] Canada's population grows from 32.5 million in 2006 to 41.8 million in 2050, while China's population starts at 1.31 billion in 2006, peaks at 1.46 billion in 2030, and then gradually falls to 1.41 billion in 2050. Canada's economy grows relatively rapidly for an OECD country (about 2.0% per year from 2006 to 2050), but not nearly as fast as China's during the entire modelling period.[5] Canada's status as a net energy exporter expands over time due to growing oil sands development in Alberta; China's energy requirements are increasingly met by oil imports, domestic abundant coal and nuclear power. Both countries experience improvements in energy efficiency, with China's overall efficiency improving dramatically as its capital stock is built and rebuilt to levels approaching international best practices.

3.2. Carbon price: Global price of $10/tCO$_2$e in 2013 rising to $100/tCO$_2$e in 2050

In our general Carbon price scenario, the same carbon price path, $10/tCO$_2$e in 2013 rising exponentially to $100/tCO$_2$e, in 2050, is applied to both countries.

3.3. Carbon-plus: Global price plus additional measures to achieve a LCS

In the Carbon-plus scenario, we impose additional measures for both China and Canada to achieve an LCS target. These measures are distinctly different, however, based on the national circumstances of each country.

For CIMS-Canada we impose sufficient additional measures to explicitly achieve a 50% emissions reduction by 2050. One of CIMS' key strengths is the ability to apply economy wide pricing policy *and* technology-specific regulations. To demonstrate CIMS' capabilities in this regard, in addition to doubling the carbon price of the last scenario, we have imposed increasingly stringent codes on building and freight rolling-stock, and an aggressive vehicle GHG emissions standard on the personal motor fleet.[6]

For China, based on the principle that developing countries' first priority is to improve the economic standard of living of their citizens while meeting local and global environmental goals, we chose emissions stabilization between 2010 and 2050 as the LCS target. This is an extremely ambitious target for an economy expected to grow more than *ten times* larger by 2050, and is achieved by applying an additional package of aggressive regulations to the Carbon price scenario.

4. Results

4.1. The Base case, Carbon price and Carbon-plus scenarios in Canada

Figure 1 and Table 2 illustrate the results of our baseline and several GHG policy simulations in Canada. The top line in Figure 1 illustrates a forecast of Canada's energy-related emissions from 2005 through 2050, including both combustion and process categories (but not inclusive of agriculture, waste and land-use change), which almost doubles from 662 Mt to 1,177 Mt in the absence of significant GHG policy. SO$_x$ emissions rise 110% in the Base case scenario, while NO$_x$ emissions rise 23%, both inclusive of all planned regulations. The next line down illustrates the effect of applying the common Carbon price schedule, which causes a decrease in emissions of

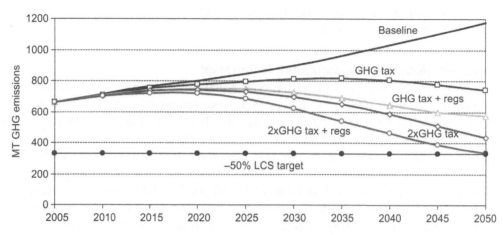

FIGURE 1 Base case, Carbon price and Carbon-plus scenarios in Canada.

TABLE 2 GHG emissions in Canada (MtCO$_2$e) in BAU and in response to carbon pricing and regulations, including percentage loss in GDP compared with BAU and SO$_x$ and NO$_x$ co-benefits

	2005	2015	2020	2025	2030	2040	2050	%Δ GHGs 2050 vs. 2005	%Δ SO$_x$ 2050 vs. 2005	%Δ NO$_x$ 2050 vs. 2005
BAU	662	765	801	846	896	1,031	1,177	+78%	+110%	+23%
GHG price	662	751	776	796	809	807	742	+12%	+26%	+4%
Price + regulations	662	736	748	748	722	645	574	−13%	+12%	−3%
2 × price	662	734	743	728	697	588	437	−4%	−20%	−8%
2 × price + regulations	662	719	718	684	619	461	334	−50%	−26%	−12%

435 Mt/year to 742 Mt in 2050; emissions in 2050 are still slightly above today's levels, primarily due to Canada's steadily growing economy, population and oil sands industry, which will remain globally competitive unless there are radical changes in use of transportation fuels or a collapse in oil prices.[7] When we add buildings and transport regulations to correct market failures in those sectors, emissions drop another 168 Mt in 2050 to 574 Mt, about 15% less than today's total, but still far short of the LCS goal of 50% reduction from today's levels. To meet the LCS target, we need to double the Carbon price schedule and apply complementary regulations in transport and buildings, which eventually reduces Canada's 2050 emissions to 334 Mt.

Apart from deep GHG abatement, the policy simulation with the doubled carbon price also leads to co-benefits of significant SO$_x$ and NO$_x$ abatement. SO$_x$ and NO$_x$ emissions fall because of an accelerated switching to low- or zero-emissions electricity sources (e.g. wind) and away from

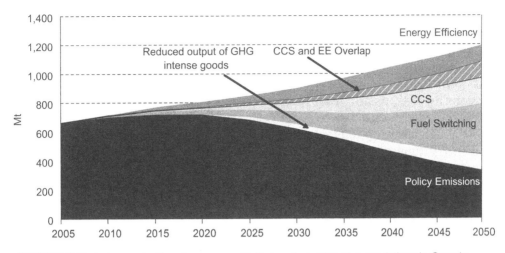

FIGURE 2 Emissions reductions by source with 2 × carbon price plus regulations in Canada.

coal, RPP and natural-gas-driven sources. The application of CCS also eliminates most SO_x and PM emissions, as well as NO_x with certain types of CCS (IPCC, 2005).

Figure 2 breaks down Canada's GHG emissions between 2005 and 2050 in terms of actions to reduce emissions. The top line represents Canada's baseline GHG emissions forecast. The top wedge marked 'energy efficiency' and the wedge below marked 'CCS and EE overlap' collectively represent the contribution of energy efficiency. The 'CCS and EE overlap' wedge and the lighter grey wedge marked 'CCS' collectively represent the contribution of CCS. The striped wedge represents the energy efficiency penalty of CCS. The darker grey wedge marked 'fuel switching' represents the contribution of switching to less carbon-intensive energy forms. The thin light-grey wedge represents emissions reduction due to 'reduced output of GHG-intensive goods' and the solid black area represents the remaining GHG emissions.

Sector emissions reductions varied widely on an absolute basis but not relatively compared with their business-as-usual (BAU) emissions. Industry, buildings, transportation and energy supply categories all reduced their emissions between 56% and 79% (Table 3).

TABLE 3 Baseline and policy sectoral emissions (Mt)

	Baseline emissions	2 × price + regulations policy	Reductions	Reductions as % of BAU	Reductions as % of total reductions
Industry	193	69	124	64%	15%
Buildings	121	53	68	56%	8%
Transportation	272	86	186	68%	22%
Energy supply	592	126	466	79%	55%
Total	1,177	334	843	–	100%

- The energy supply sector, including electricity generation, oil and gas extraction and processing, coal mining, and refining, achieves the greatest reduction in both absolute and relative terms, reducing emissions 79% compared with the 2050 baseline (55% of total emissions reductions). These reductions came primarily from fuel switching and CCS.
- Buildings contributed the least overall emissions reductions (8%), but the sector still reduced its emissions by 56%, mainly from fuel switching and increased efficiency.
- Transportation reduced its emissions considerably (68%), contributing 22% of overall reductions, primarily through efficiency improvements and some switching to biofuels and hydrogen.
- The non-energy supply industry reduced its emissions by 64%, via efficiency improvements, fuel switching, CCS and some output reduction in cement and chemical production.

4.2. The Base case and Carbon price scenarios in China

While energy and GHG emissions data in developed countries such as Canada are relatively transparent, the same does not necessarily hold for developing nations such as China, where the difficulties begin with ascertaining current emissions levels. With the intention of reducing environmental degradation and improving the safety record of the coal industry, in 1997 the Chinese central government launched a campaign to close small, private coalmines, and imposed a stringent national production cap on the coal industry. However, lured by the tax revenue, local governments in coal-mining regions were unwilling to comply with the central government's production cap and simply underreported the coal supply and demand within their geographical boundaries, and many of the small coal mines that were ordered to be shut down stayed open. Liu Xueyi and Liu Anhua estimated that the amount of unreported coal production was 52 million, 225 million, 260 million, 175 million and 173 million tonnes for the years 1998, 1999, 2000, 2001 and 2002, respectively (Tu, 2007). Although the National Bureau of Statistics (NBS) recently revised China's energy statistics, there are still many inconsistencies in official statistics for 2000. As a result, 2005 instead of 2000 is deliberately selected as the baseline year of our study. While the emissions coefficients used in this study are consistent with the Chinese government's *Initial National Communication on Climate Change* submitted to the UNFCCC, they are generally lower than the default values in the 2006 IPCC *Guidelines for National Greenhouse Gas Inventories*. As a result, China's base-year emissions in this study probably represent a lower range estimate of China's actual emissions levels in 2005. Table 4 illustrates the reference simulation results of China's energy sector. Between 2005 and 2050, primary energy consumption in China is expected to grow from 2,230 million tonnes of coal equivalent (1 Mtce = 29.31 PJ) to 5,841 Mtce. Although the development of natural gas infrastructure and renewables is projected to grow quickly, coal still represents more than 50% of China's primary consumption in 2050, compared with 23% from oil, 11% from natural gas, and 14% from primary electricity.

By 2050, more than 80% of China's potential hydro resources are expected to be exploited, accounting for 14.6% of electricity output. The shares of nuclear and renewables in the electricity generation mixture rise to 12.4% and 1.5%, respectively, in 2050. In comparison, the percentage of fossil fuel thermal power output declines over time to 71.5% in 2050. Moreover, the efficiency of fossil fuel thermal power increases from 35.1% in 2005 to 45.7% in 2050. China's GHG intensity per $GDP declines over time, reaching 0.71 $kgCO_2e/\$$ in 2050, an 81% reduction compared with 2005 levels. However, China's absolute GHG emissions still grow about 2% annually until 2050, and per capita emissions are expected to increase from 4.5 tCO_2e/cap in 2005 to nearly 10 tCO_2e/cap in 2050, largely due to increased personal mobility and consumption of consumer goods.

TABLE 4 Reference scenario of China's energy sector, 2005–2050

	2005	2010	2020	2030	2050
Final energy consumption (Mtce)	1,579	1,941	2,548	3,130	4,086
Agriculture (%)	3.2	3.1	3.0	3.0	2.8
Industry (%)	69.3	69.1	66.3	62.7	55.4
Transportation (%)	14.0	14.2	16.0	17.8	20.5
Commercial (%)	3.4	3.7	4.1	4.8	7.3
Residential (%)	10.1	9.9	10.6	11.7	14.0
Primary energy consumption (Mtce)	2,230	2,789	3,674	4,506	5,841
Coal (%)	69.5	66.7	63.3	59.9	52.6
Oil (%)	20.6	21.4	21.3	21.8	23.1
Gas (%)	2.9	3.9	5.9	7.6	10.8
Hydro (%)	6.2	6.7	6.6	6.6	6.6
Nuclear (%)	0.8	1.2	2.5	3.5	5.7
Other new and renewable (%)	0.0	0.2	0.4	0.7	1.2
Electricity output (TWh)	2,502	3,278	4,762	6,419	9,954
Coal (%)	78.4	76.3	72.0	69.7	64.9
Oil (%)	2.8	1.8	1.3	0.8	0.6
Gas (%)	0.7	1.7	3.9	4.8	6.0
Hydro (%)	15.9	16.9	16.0	15.5	14.6
Nuclear (%)	2.1	3.0	6.0	8.2	12.4
Other new and renewable (%)	0.1	0.3	0.8	1.0	1.5
Carbon emission (MtCO$_2$e)	5,933	7,412	9,514	11,370	13,903
Fuel combustion	85.1	83.9	84.2	84.6	85.6
Industrial process[a]	10.3	11.7	11.5	11.2	10.6
Fugitive[b]	4.5	4.4	4.3	4.1	3.8
Per capita carbon emission (tCO$_2$e/capita)	4.54	5.48	6.69	7.80	9.87
Energy intensity per GDP (Kgce/US$, 2000 price)	0.97	0.76	0.53	0.38	0.21
Carbon intensity/energy consumption (tCO$_2$e/tce)	3.76	3.82	3.73	3.63	3.40
Carbon intensity/GDP (KgCO$_2$e/US$, 2000 price)	3.66	2.91	1.99	1.39	0.71

[a] Process emissions of cement, lime, iron and steel, and calcium carbide production.
[b] Fugitive emissions of the coal mining, and upstream oil and gas industry.

During the climate change negotiation process in the 1990s, China teamed with most developing countries to successfully impede any binding emission reduction initiatives for developing countries. Since 2000, China's spike in coal and oil use has pushed its GHG emissions to a record high, and the Chinese government is facing increasing pressure from the international community for legally binding climate commitments. Nevertheless, China's recent involvement in the Asia–Pacific

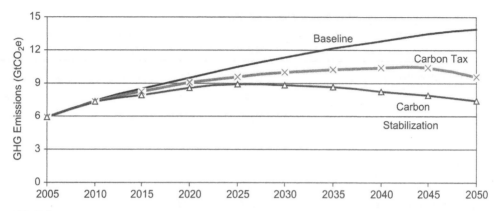

FIGURE 3 Base case, Carbon tax and Carbon stabilization scenarios in China.

Partnership on Clean and Development and Climate and its first Energy White Paper all indicate that while Beijing will continuously reject any request for legally binding climate change commitment, it is likely to adopt an intensity-based approach to lower its GHG emissions trajectory in the future (e.g. China aims to reduce its energy intensity by 20% during the 11th five-year plan period; 2005–2010).

While China's attempt to enforce intensity-based measures is an important first step in the right direction, minimizing the risk of global climate disruption will ultimately require absolute emissions reductions from major developing countries. For instance, in the reference scenario, while China's GHG emissions intensity drops 79% between 2005 and 2050, China's total GHG emissions still increase 159%. In Figure 3, we illustrate two GHG policy simulations, the first where China participates in the global Carbon price scenario, and the second where China implements supplementary regulations sufficient to stabilize emissions at 2010 levels by 2050.

The Carbon price scenario results in a 31% emissions reduction in 2050, but China's 2050 GHG emissions under the Carbon price scenario are still 61% higher than 2005 levels, suggesting that stronger measures may be necessary. Considering that China has already been exempted from binding emissions reduction commitments during the Kyoto Protocol period (1990–2012), the 'Carbon stabilization' line in Figure 3 uses 2010 as China's baseline year for GHG emissions reduction. Addition of complementary regulations to the Carbon price schedule generates a 47% emissions reduction in 2050 compared with the reference simulation, returning emissions to 2010 levels. The complementary regulations include:

- accelerated closure of small power plants in the electricity industry
- subsidies for renewables
- accelerated de-commission of inefficient heavy industrial plants
- vehicle efficiency standards
- voluntary initiatives/public environmental campaign.

Figure 4 describes the sources of emissions reduction in the *Carbon stabilization* scenario. Energy efficiency contributes more than 25% of emissions reductions in 2050, while CCS is another significant source, representing 25% of GHG emissions reductions at the end of the modelling period. However, due to the energy penalty imposed by CCS, some of the gains from energy conservation are lost. Fuel switching contributes 21% of emissions reductions in 2050.

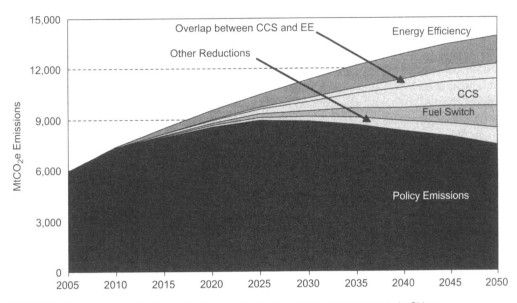

FIGURE 4 Sources of emissions reductions in the Carbon stabilization scenario in China.

Table 5 describes the emissions reductions by sector.

▓ Similar to Canada, the energy supply sector in China also achieves the highest percentage emissions reduction in both absolute and relative terms, lowering its emissions 61% compared with the 2050 baseline (60% of total emissions reductions). These reductions came primarily from fuel switching and CCS.
▓ The industrial sector contributes 27% of total GHG emissions reductions in 2050. These are from energy efficiency, fuel switching and reduced output of GHG-intensive goods.
▓ Transportation also reduces its emissions relative to the baseline (24%; 6.3% of overall reductions), primarily through efficiency standards with some fuel switching.

TABLE 5 Baseline and policy sectoral emissions (MtCO$_2$e)

	Baseline emissions	Stabilization emissions	Reductions	Reductions as % of BAU	Reductions as % of total reductions
Industry	4,651	2,879	1,772	38%	27%
Commercial	249	158	91	37%	1.4%
Residential	569	380	188	33%	2.9%
Transportation	1676	1,270	406	24%	6.3%
Agriculture	165	116	49	30%	0.8%
Energy supply	4,591	2,622	3,972	60%	61%
Total	13,903	7,425	6,478		100%

If China's development path unfolds as portrayed in our reference scenario, China's ambitious economic growth over the next five decades will come with the price of significant climate change risks, deteriorating air quality and greater energy security vulnerability. Our analysis suggests that it is plausible to employ appropriate policy instruments to induce an alternative development path which would enable China to continue social and economic development while reducing global climate change risks, and the keys to a successful carbon stabilization strategy for developing countries such as China are: (1) to set a politically acceptable baseline year (e.g. 2010 instead of 1994); (2) to take early action (e.g. impose a carbon price soon, e.g. 2013); and (3) to allow a relatively longer period to achieve a quantity-based emissions reduction target (compared with the short time span of the Kyoto Protocol).

4.3. Interactions between China and Canada

We now compare the marginal GHG abatement costs of CIMS-China and CIMS-Canada, and discuss what they imply about potential carbon permit flows. Table 6 shows the responses of CIMS-Canada and CIMS-China to constant carbon prices from 2013 through 2050.[8] We used these pricing schedules to calculate an approximate response of the developed world and China to a combined 50% reduction target from 2005 emissions levels. We found these regions could collectively reach the target at about $175/tCO$_2$e, with permits flowing from the developed countries to China.

TABLE 6 China and Canada's 2050 marginal abatement schedules (GtCO$_2$e)

| Carbon price ($/tCO$_2$e) | 0 | 10 | 25 | ← China cheaper | | Developed countries cheaper → | | | |
				50	75	100	125	150	175
China's 2050 BAU	13.9								
China's reductions		1.2	2.6	4.2	5.1	5.8	6.4	6.8	7.1
% of 2050 BAU		8%	19%	30%	37%	42%	46%	49%	51%
Canada's 2050 BAU	1.2								
Canada's reductions		0.1	0.1	0.3	0.4	0.6	0.7	0.8	0.8
% of 2050 BAU		5%	10%	24%	35%	51%	61%	67%	71%
Developed Country 2050 BAU	18.7								
Developed Country Reductions		0.9	1.9	4.5	6.6	9.4	11.3	12.5	13.3
China+Developed Countries 2005 BAU	24.2								
China+Developed Countries reductions		2.1	4.5	8.7	11.7	15.3	17.7	19.3	20.4
% World emissions in 2050 compared with 2005		126%	116%	99%	86%	72%	62%	55%	51%

Reaching this conclusion required assembling several calculations and assumptions. The baseline emissions for Canada and China are from CIMS, as are the abatement schedules. The developed world baseline is calculated by multiplying the United Nations Secretariat Population Division developed-country 2050 population forecast (1.245 billion) by an average GHG intensity of 15.0 tCO_2e per capita.[9] The developed-world reductions are calculated by extrapolating the Canadian abatement schedule to the baseline 2050 developed-world emissions forecast, a debatable assumption that is subject to further research.

An interesting finding from this exercise is that, while relative reductions costs are lower in China up to $75/t, at higher prices reductions are relatively less costly in the developed world. A key reason for this is the immense economic growth predicted for China. Many Chinese people still lack the basic energy and services taken as a given in the developed world. No matter how energy- or GHG-efficient these services may be when finally provided, if they are not carbon-free they will almost unavoidably add to total GHG emissions.

5. Discussion

In this article, using the CIMS hybrid technology simulation model, we have provided Chinese and Canadian baseline GHG emissions forecasts to 2050, their responses to a global carbon price of $10t/CO_2e$ in 2013 rising to $100t/CO_2e$ by 2050, and alternative policies to achieve a 50% reduction in GHG emissions compared with 2005 in the case of Canada and stabilization at 2010 emissions levels for China. Our results indicate that a global low-carbon society is feasible, with the condition that policy makers are willing and able to impose long-term, credible policy packages with carbon pricing policy as the core element, coupled with supplementary regulations to address market failures.

In terms of actions to reduce emissions, although energy efficiency could play an important role in both Canada and China, it is still only part of the solution. The scale of the LCS challenge requires that all available GHG emissions reduction options be vigorously pursued, including fuel switching to lower- or zero-carbon fuels, CCS, and accelerated development of renewables. People, especially those living in developed countries, should also expect to lower their consumption of inherently GHG-intensive products.

This study differs from many in the technology simulation literature in that the modelling framework used to run the simulations is not an optimization model, but is instead designed to simulate the investment and consumption responses of firms and consumers as accurately as possible. In this regard, the model is less concerned with least-cost solutions as opposed to measuring the stringency of policies necessary to achieve specific levels of real emissions reductions. Our results may be seen as being inclined towards a perspective that does not count on fundamental behavioural change to achieve deep reductions; e.g. we assume that people respond to the cost of air travel, not to the ethical questions posed by its associated pollution. If the general public, especially in developed nations, could be convinced to radically change their GHG emissions patterns, e.g. to voluntarily reduce travel and consumption of GHG-intensive goods despite rising incomes, then a lower carbon price estimate than we provide may achieve the LCS target. The subject of radical behaviour change, and how to induce it for public good reasons while respecting overall welfare and liberty, is of great importance to the climate change debate and policy, but beyond the scope of our analysis.

This study also suggests that the roles of the developing and developed countries will be distinctly different. Developing countries such as China cannot be expected to voluntarily sacrifice opportunities for economic development in exchange for global environmental quality, given the historical free use of the atmosphere by developed nations. Assuming that global emissions targets will become gradually more stringent over time, with an equivalent rise in carbon prices, our analysis indicates that the developing world will initially provide the least cost opportunities for emissions reduction, followed in later years by the developed countries. This dynamic relationship argues for flexible policy instruments, for instance emissions cap-and-trade systems, which can adapt as circumstances change. In the near future, this offers the developed world a tremendous opportunity to meet their climate change commitments (e.g. through the Clean Development Mechanism). In the long run, the capacity discrepancy between developed and developing countries, their differing national priorities, and the global commons nature of climate risks, indicate that the developed countries will probably be required to subsidize low-GHG development initiatives in the developing world (e.g. through knowledge transfer and voluntary assumption of relatively tighter emissions caps) while reducing their own emissions, with the eventual goal of a global GHG emissions cap, dictated by the maximum allowable atmospheric GHG concentration. If flexible carbon pricing policies and knowledge transfer, supplemented by regulations to address market failure and adjusted according to the capacity and needs of the developed and developing countries, can be effectively embedded into future developments of the UNFCCC or similar international agreements, a window of opportunity may open for developing countries to directly participate in the global LCS initiative.

Notes

1. For a summary of global climate change science, see the Intergovernmental Panel on Climate Change (IPCC, 2007).
2. See the Synthesis Report (IPCC, 2007) of the IPCC Fourth Assessment Report.
3. Canada's national energy use and GHG forecast, *Canada's Energy Outlook 2006*, is prepared by Natural Resources Canada. It provides a forecast for all energy use and GHG emissions out to 2020. For 2020–2030, we have used a forecast prepared by Informetrica Ltd., the leading Canadian economic forecasting company.
4. The reference scenario of Tu et al. (2007) only covers the period between 1995 and 2030. In this study, the coverage of the Base case has been extended to 2050, taking China's recent energy statistics revision into consideration.
5. We assume that China could exceed the GDP quadrupling target between 2000 and 2020. After 2020, China will grow at a gradually slower pace, but the average annual GDP growth rate still reaches 4.8% between 2020 and 2050.
6. These regulations were developed as part of a study for the Canadian National Round Table of the Environment and the Economy, a study designed to look at the feasibility and costs of reducing Canada's GHG emissions by –45%, –65% and –80% from 2005 levels by 2050 (Bataille et al., 2007a).
7. Canada's oil sands industry is projected to grow an average of 5–7% per year to 2020 according to NRCan, after which growth is projected to slow to 3% per year. Based on our assumptions of export demand for crude oil, we have cut post-2020 growth to 1.5% per year for the Carbon price scenario, and to 0% per year in the Carbon-plus scenario. We do not reduce production after 2020, as we have assumed oil demand and prices will sustain operating costs, if not the cost of amortizing new production.
8. China currently carries out half the world's planned CDM projects, most related to CFC reduction. These are assumed to occur in the baseline scenario.
9. The population forecast is from United Nations Secretariat Population Division (2007). The intensity is an average between the European G-7 and the North American G-7. See Bataille et al. (2007b) for details.

References

Bataille, C., Jaccard, M., Nyboer, J., Rivers, N., 2006, 'Towards general equilibrium in a technology-rich model with empirically estimated behavioral parameters', *Energy Journal* 27 (Special Issue: *Hybrid Modeling of Energy–Environment Policies: Reconciling Bottom-up and Top-Down*).

Bataille, C., Rivers, N., Peters, J., Tu, J., 2007a, *Pathways and Policies for Long-term Greenhouse Gas and Air Pollutant Emissions Reductions*, Contract #NRT 2007060, Prepared for the National Round Table on the Environment and the Economy by J. & C. Nyboer and Associates, 6 July 2007.

Bataille, C., Rivers, N., Mau, P., Joseph, C., Tu, J., 2007b, 'How malleable are the greenhouse gas emission intensities of the G7 nations?', *Energy Journal* 28(1), 145–170.

Government of China, 2004, *The People's Republic of China Initial National Communication on Climate Change*, a submission prepared for the United Nations Framework Convention on Climate Change, October 2004.

Hourcade, J.-C., Jaccard, M., Bataille, C., Ghersi, F., 2006, 'Hybrid modeling: new answers to old challenges', *Energy Journal* 27 (Special Issue: *Hybrid Modeling of Energy–Environment Policies: Reconciling Bottom-up and Top-Down*), 1–12.

IPCC (Intergovernmental Panel on Climate Change), 2005, *Special Report on Carbon Dioxide Capture and Storage*, prepared by Working Group III of the Intergovernmental Panel on Climate Change, B. Metz, O. Davidson, H.C. de Coninck, M. Loos, L.A. Meyer (eds). Cambridge University Press, Cambridge and New York, 442 pp.

IPCC (Intergovernmental Panel on Climate Change), 2006. *2006 IPCC Guidelines for National Greenhouse Gas Inventories Programme*, in: H.S. Eggleston, L. Buendia, K. Miwa, T. Ngara, K. Tanabe (eds), IGES, Japan.

IPCC (Intergovernmental Panel on Climate Change), 2007, *Climate Change 2007: Synthesis Report. Contribution of Working Groups I, II and III to the Fourth Assessment Report of the Intergovernmental Panel on Climate Change*, Core Writing Team: R.K. Pachauri, A. Reisinger (eds), IPCC, Geneva, 104 pp.

Liu, X., Liu, A., 2004, 'Estimation and recommendations regarding production levels of township and village coal mines and national coal output in recent years', *Energy Policy Research* 1, 38–42.

Natural Resources Canada, 2006, *Canada's Energy Outlook: The Reference Case 2006*, Natural Resources Canada [available at www.nrcan.gc.ca/com/resoress/publications/peo/peo-eng.php].

Population Division of the Department of Economic and Social Affairs of the United Nations Secretariat, 2007, *World Population Prospects: The 2006 Revision* and *World Urbanization Prospects: The 2005 Revision* [available at http://esa.un.org/unpp].

Stavins, R., 1998, 'What can we learn from the grand policy experiment? Lessons from SO_2 allowance trading', *Journal of Economic Perspectives* 12(3), 69–88.

Stavins, R., Jaffe, J., Schatzki, T., 2007, *Too Good to Be True? An Examination of Three Economic Assessments of California Climate Change Policy*, AEI–Brookings Joint Center for Regulatory Studies.

Tu, J., 2007, 'China's botched coal statistics', *China Brief* 6(21).

Tu, J., Jaccard, M., Nyboer, J., 2007, 'The Application of a hybrid energy-economy model to a key developing country: China', *Energy for Sustainable Development* 11(1), 35–47.

■ research article

Back-casting analysis for 70% emission reduction in Japan by 2050

JUNICHI FUJINO[1]*, GO HIBINO[2], TOMOKI EHARA[2], YUZURU MATSUOKA[3], TOSHIHIKO MASUI[1], MIKIKO KAINUMA[1]

[1] National Institute for Environmental Studies, Tsukuba, Japan
[2] Mizuho Information & Research Institute, Inc., Tokyo, Japan
[3] Kyoto University, Kyoto, Japan

This article envisions a future in which advances in technology and urban development have transformed Japanese society by 2050, resulting in significant greenhouse gas reductions. Pathways leading Japan towards a low-carbon society are calculated using a scenario approach based on 'back-casting' techniques. It is possible to reach a 70% reduction in CO_2 emissions through a combination of demand-side and supply-side actions. On the demand side, reductions of 40–45% are possible through efficiency improvements, decreased population and the more rational use of energy despite increased energy demands arising in certain sectors. On the supply side, CO_2 emissions can be reduced through a combination of the appropriate choice of low-carbon energy sources (including carbon capture and storage) and improving energy efficiency. The estimated direct annual cost of technology to achieve this by 2050 is 6.7–9.8 trillion yen, approximately 1% of the estimated 2050 GDP. However, this excludes costs involved in infrastructure investments with aims other than climate policy (e.g. strengthening international competitiveness, improving security, enhancing urban development, and reinforcing energy). To avoid investing in its current high-carbon-emitting infrastructure, Japan must develop long-term strategies to create the necessary technological and societal innovations and to channel the appropriate financial resources for intensive economy-wide change, such as development of land, urban areas, and buildings, improvements in industrial structures, and new technologies.

Keywords: back-casting; climate change; emission reduction; energy; innovation; Japan; low-carbon society

Ce papier conçoit un futur dans lequel les avancées technologiques et l'urbanisation auraient transformé la société japonaise d'ici 2050, entraînant de fortes réductions en gaz à effet de serre. Les itinéraires entraînant le Japon vers une société sobre en carbone sont calculés selon une approche de scénarios basée sur les techniques de rétroprojection. Il serait possible d'obtenir un taux de réduction de 70% des émissions de CO_2 en alliant des actions du côté de la demande ainsi que de l'offre. Du côté de la demande, des réductions de l'ordre de 40–45% seraient possible par des progrès dans l'efficacité, une réduction de la population et une utilisation plus rationnelle de l'énergie, malgré une demande en énergie plus accrue dans certains secteurs. Du côté de l'offre, les émissions de CO_2 pourraient être réduites par une combinaison du choix approprie de sources d'énergie sobres en carbone (comprenant la capture et le stockage du carbone) et l'amélioration de l'efficacité énergétique. Le coût technologique annuel pour y arriver d'ici 2050 est estimé à 6.7–9.8 trillions de yen, soit environ 1% du PIB estimé pour 2050. Cependant, ceci exclut le coût impliqué pour les investissements dans l'infrastructure à but autre que la politique climatique (par exemple le renforcement de la compétitivité internationale, l'amélioration de la sécurité, l'amélioration urbaine, et le renforcement énergétique). Pour éviter d'investir dans l'infrastructure actuelle hautement émitrice en carbone, le Japon devra développer des stratégies à long terme pour créer les innovations technologiques et sociétales nécessaires pour canaliser les ressources financières appropriées de manière a encourager les changements à travers l'économie, tels

■ *Corresponding author. E-mail*: fuji@nies.go.jp

CLIMATE POLICY 8 (2008) S108–S124

doi:10.3763/cpol.2007.0491 © 2008 Earthscan ISSN: 1469-3062 (print), 1752-7457 (online) www.climatepolicy.com

que dans l'exploitation des terres, les régions urbaines, les bâtiments, l'amélioration des structures industrielles, et les nouvelles technologies.

Mots clés: changement climatique; énergie; innovation; Japon; réduction des émissions; rétroprojection; société sobre en carbone

1. Introduction

One important characteristic of the climate system is its inertia. Because of past and current greenhouse gas (GHG) emissions, some increase in global temperature is unavoidable. Such increases in temperature carry profound risks. Even a small increase in temperature is likely to have significant impacts on ecosystems and species, and might lead to increased drought and extreme rainfall events, with severe consequences for our society.

In the IPCC Fourth Assessment Report, released in 2007 (IPCC, 2007a), the limit of increase in average worldwide temperatures is suggested as 2–3°C above the current level. In the WGIII report (IPCC, 2007b), to stabilize temperature rise below 2.6°C above pre-industrial level needs 30–60% CO_2 emission reductions by the year 2050 as compared to 2000.

On 24 May 2007, former Prime Minister Shinzo Abe gave an after-dinner speech at the Future of Asia International Conference entitled '*Invitation to 'Cool Earth 50' – Three Proposals, Three Principles*'. The speech covered (1) a proposal for a long-term target of cutting global emissions by half from the current level by 2050 as a common goal for the entire world, (2) three principles for establishing an international framework to address global warming from 2013 onwards, and (3) the launch of a national campaign for achieving the Kyoto Protocol target. In particular, the first proposal for cutting emissions by half included the specific suggestions of developing innovative technologies and building a low-carbon society (LCS). Furthermore, at the G8 Summit held at the beginning of June this year at Heiligendamm, Germany, there was agreement among the leaders to give serious consideration to reductions of at least 50% by the year 2050.

For the per capita emission in 2050 to be the same across the world, this would entail Japan reducing its emissions by about 80% compared with the 1990 level. However these numbers include a certain amount of uncertainty arising from global warming mechanisms and climate impacts. A large amount of reduction is required, considering the balance of sinks and sources of GHG. This implies that the reduction rate for Japan would have to be in the range of 60–80%.

The current policy context in Japan is best summarized by the three proposals that the Government of Japan has decided to implement, as explicated in Prime Minister Yasuo Fukuda's address as the Chair of the G8 Summit on the occasion of the Annual Meeting of the World Economic Forum in Davos, Switzerland (Fukuda, 2008). First, the government has proposed the equity of reduction obligations and a bottom-up approach to set sector-wise targets based on the potential for efficiency improvement and technologies in the future years. Second, Japan will establish a new financial mechanism, the 'Cool Earth Partnership', on a scale of US$10 billion, to cooperate with the developing countries' efforts to reduce emissions. Third, in order to halve greenhouse emissions by 2050, Japan has proposed to invest about US$30 billion in research and development in the fields of energy and environment with the objective of promoting technological innovations.

This study analyses the possibility of achieving a LCS in Japan, where CO_2 emissions, one of the major greenhouse drivers, would undergo a 70% reduction by 2050 below the 1990 level. The study is part of the research project titled 'Japan Low-Carbon Society Scenarios towards 2050' undertaken during 2004–2008. About 60 researchers in Japan extensively studied the potential for carbon reduction in different sectors such as transportation, information and communication, households, industry, energy, etc. The project's methodology and outcomes have been presented at several conferences and published in the interim report *Aligning Climate Change and Sustainability* (Matsuoka et al., 2007).

Transformation in social, economic and technological activities is expected during the first half of the century. The range of such transformation varies widely. It is necessary to make preparations for the desired socio-economic changes to achieve LCS.

Assuming that such a degree of socio-economic change is possible, the back-casting method was adopted in this study to examine the strategies for achieving the LCS. Some of the key aspects of this method are shown in Figure 1. Among the most important steps of this process we could highlight the following:

- To envision the direction of future Japanese socio-economic structure towards 2050 within a certain range (for instance, *Scenario A*: active, quick-changing and technology-oriented, and *Scenario B*: calmer, slower and nature-oriented) and to describe the characteristics of those two types of societies qualitatively.
- To quantify the behaviour of people and households (how people spend time, what services will be needed), design of city and transportation (what kinds of city and houses people live in, how people travel), and industrial structure (estimation of the structural changes by a multi-sector computable general equilibrium model) for each scenario, and to estimate energy-service demand for each scenario (for instance, the volume of cooling, hot water supply, crude steel production, and transportation demand).
- To calculate energy services demand, while satisfying the CO_2 emission reduction target that supports the estimated socio-economic activity in each scenario; to explore the appropriate combination of energy services demand, end-use energy technology (air conditioner, thermal insulation, boiler, steel plant, hybrid car, etc.), types of energy supply and energy supply technologies, based on the consideration of the available volume of energy supply, its cost-efficiency and its political feasibility; to identify the types of energy demand and supply technologies as well as their shares.
- To quantify the primary and secondary energy demands and the amount of resulting CO_2 emissions.

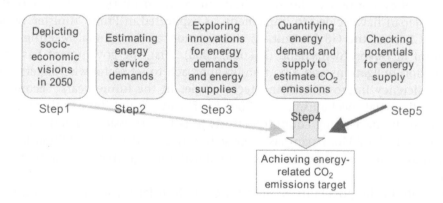

FIGURE 1 LCS scenario approach.

2. Narrative scenarios for Japan in the 2050s

Possible pictures of Japan in the future and their pathways towards the LCS have been discussed by experts in various fields. Two scenarios, A and B, have been developed as a result (Table 1; see also Step 1 in Figure 1). The features of *Scenario A* can be described as an active, quick-changing and technology-oriented society. On the other hand, *Scenario B* is a calmer, slower and nature-oriented society. Changes in social indicators and various assumptions made in both scenarios are within the ranges of existing major studies of Japanese future society projections (Cabinet Office, 2005). In reality, however, future Japanese society may be a mixture of elements from both the scenarios.

In scenarios A and B, the annual growth rate of per capita GDP has been assumed to be 2% and 1%, respectively. The changes in services demand that are directly related to energy consumption (e.g. heating, transportation, and built environment management) have been set by assuming the changes in lifestyle of the representative people in each scenario. The improvement in service levels has been assumed to be gradual and moderate. Excessive service demands such as heavy loads in households and offices (such as air-conditioning throughout the day), or a disorganized urban structure with inappropriate urban planning which leads to considerable increase in traffic demand, is not assumed in the scenarios.

TABLE 1 Keywords of the two scenarios

Keywords	Scenario A	Scenario B
Mindset of people		
Goal of life	• Social success	• Social contribution
Residence	• Urban orientation	• Rural orientation
Family	• Self-dependent	• Cohabitation
Acceptance of Advanced technology	• Positive	• Prudent
Population		
Birth rate	• Downslide	• Recover
Immigration of foreign workers	• Positively accepted	• Status quo
Emigration	• Increase	• Status quo
Landuse and cities		
Migration	• Centralization in large cities	• Decentralisation
Urban area	• Concentration in city centre	• Population decrease
	• Intensive land use in orban area	• Maintain minimum city function
Countryside	• Significant population decrease	• Gradual population decrease
	• Advent of new businesses for efficient use of land space	• Local town development by local communities & citizens

Keywords	Scenario A	Scenario B
Life and household		
Work	• Increase in "Professionals"	• Work sharing
	• High-income & over-worked	• Working time reduction & equalization
Housework	• Housekeeping robots & services	• Cooperation with family & neighbours
Free time	• Paid-for activity	• With family
	• Improving carrier	• Hobby
	• Skill development	• Social activity (i.e Volunteer activity)
Housing	• Multi-dwellings	• Detached houses
Consumption	• Rapid replacement cycle of commodities	• Long lifetime cycle of commodities (Mottainal)
Economy		
Growth rate	• Per capita GDP growth rate: 2%	• Per capita GDP growth rate: 1%
Technological development	• High	• Not as high as scenario A
Industry		
Market	• Deregulation	• Adequate regulated rules apply
Primary Industry	• Declining GDP share	• Recovery of GDP share
	• Dependent on import products	• Revival of public interest in agriculture and forestry
Secondary Industry	• Increasing add value	• Declining GDP share
	• Shifting production sites to overseas	• high-mix low-volume production with local brand
Tertiary industry	• Increase in GDP share	• Gradual increase in GDP share
	• Improvement of productivity	• Penetration of social activity

As Scenario A is characterized by advanced technologies and individualist lifestyles, it would require special policies to promote the research and development of highly efficient technologies in key sectors such as energy, industries and transportation. On the other hand, Scenario B would require policies that induce people to change their lifestyles towards decentralized community-based units and promote communication and logistics systems to interconnect such communities. For both the scenarios, it would be necessary to estimate inventories of GHGs for each sector and activity, and appropriate levels of tax, emission trading and other mechanisms to achieve the desired reductions. Additionally, it would be crucial to plan for potential benefits such as the possibility of technological and social innovations and the consequent advantage to industrial competitiveness.

3. Estimation of service volume for each socio-economic scenario and sector

In this analysis, the population and household model (PHM), the building dynamics model (BDM), the transportation demand model (TDM), and the computable general equilibrium (CGE)

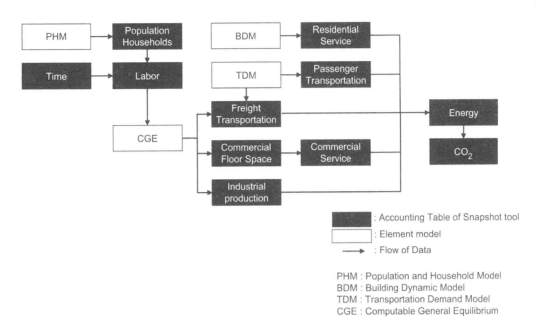

FIGURE 2 Relationship between socio-economic scenarios and tools for estimating volume of services in individual sectors.

model were used to quantify socio-economic narrative scenarios (Figure 2) and estimate the volume of services in individual sectors (Table 2) (Matsuoka et al., 2007).

1. The PHM simulates the future provincial population by age and sex, providing exogenous national and provincial base-year population, expected life tables, expected fertility rate, and expected migration rate. The number of households by family type is also calculated using the headship rate method.
2. The labour population is estimated by assuming population, employment rate, daily working hours and labour productivity.
3. The parameters in the CGE are calibrated from the input–output data in 2000. Input data include expected economic growth, labour endowment, capital endowment, technology improvement (input–output coefficient and energy efficiency), preference change, ratio of imported goods to domestic goods, international price, etc. Based on this input data, the commodity prices and activities are calculated to balance the demand and supply of each commodity.
4. The BDM uses the cohort component method to estimate the number and the floor space of future dwelling stock from the dwelling stock in the base year, the residual ratio and construction of the new dwellings. The model is able to estimate the number and the floor space of future dwelling stock by region, building type, and construction period according to the population distribution, types of new dwellings, retrofit of existing dwelling stock, and the residual ratio of dwelling stock.
5. The passenger transportation demand model (PTDM) simulates transportation demand associated with changes in population distribution, social environment, personal activity patterns, modal share, and average trip distance. It is based on the transportation demand model developed by Japan's Ministry of Land Infrastructure and Transport (MLIT). Transportation demand within the daily living area (intra-region transportation) is

TABLE 2 Assumptions of driving forces in the year 2050

		Unit	2000	2050A	2050B
Population		Million	127	94	100
Household		Million	47	43	42
GDP		Trillion (2000)	519	1,080	701
(share)	Agriculture	%	2	1	2
	Industry	%	28	18	20
	Service	%	71	80	79
Floor space		Million m²	1,654	1,934	1,718
Passenger Transportation		Billion person-km	1,399	1,034	1,080
(share)	Automobile	%	53	50	52
	Public transportation	%	34	43	42
	Walk/bike	%	7	7	6
Freight Transportation		Billion ton-km	574	500	516
Industrial production index		2000=100	100	126	90
Raw material	Steel	Million ton	107	67	58
	Ethylene	Million ton	8	5	3
	Cement	Million ton	82	51	47
	Pulp/paper	Million ton	32	18	26

calculated separately from transportation demand between the daily living areas (inter-region transportation).

6. The freight transportation demand model (FTDM) simulates the freight transportation volume associated with changes in the industrial structure, material density, transportation distance, and modal share. The inputs of the model are production and imports calculated by the CGE model. The outputs are freight transportation volumes in terms of tonne-kilometre by mode.

The industrial structure of 2050 aiming at the low-carbon target is estimated using a computational general equilibrium model (CGE) with 57 sectors. Figure 3 shows the results summed up into 40 sectors. Scenarios A and B both show the progress of the service sector, the increase in electric machines, equipment and supplies, the transport equipment industries, and the reduction in energy-intensive industries. These results do not differ from the previous estimations (*Japan's Visions for the 21st Century*, 2005, Cabinet Office). The active society (Scenario A) demonstrates remarkable progress in sectors such as commercial services and electromechanical machinery and transport machinery industries. Based on this estimation, in the following sectors, only the direct effect of measures for fulfilling the LCS is evaluated. The indirect effects such as trigger effects by the measures will not be analysed.

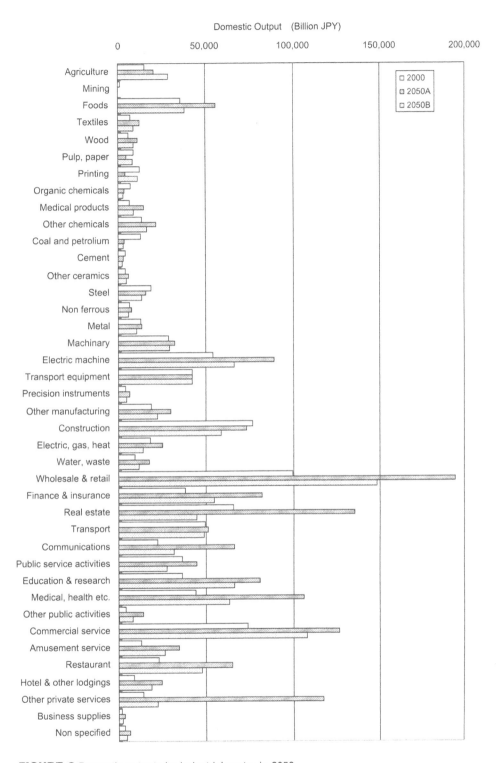

FIGURE 3 Domestic outputs by industrial sector in 2050.

4. Feasibility of achieving 70% reductions by 2050

4.1. Countermeasures to achieve LCS

Scenarios A and B select packages of technological countermeasures for the LCS on the basis of the estimated industrial structure in 2050, as shown in Figure 3. Table 3 lists key countermeasures to achieve LCS. Prospects for future technological innovations have been set, based on various reports, research articles and white papers, such as the *Strategic Technology Roadmap in Energy Field– Energy Technology Vision 2100* by the Ministry of Economy, Trade and Industry (METI), Japan (METI, 2005). After configurations of technological countermeasures, required secondary energy demands, such as grid electricity, fossil fuels, hydrogen and others, were determined (Step 3 in Figure 1).

4.1.1. Possible countermeasures to achieve LCS in the industrial sector (Figure 4)

The industrial structure would shift towards a service economy (see Figure 3). Production of electric/ transport machinery industries would also increase, enhancing industrial competitiveness of manufacturing sectors. Social infrastructures would be further developed and the volumes of steel or cement stocked in such infrastructures would increase considerably. An innovative recycling technology would be developed and the stocked materials within the society could be reused for high-quality purposes. Applications of such technologies would lead to improvements in resource usage rates of raw materials.

TABLE 3 Key technological countermeasures in the environmental option database

Sector	Technology
Residential and Commercial	Efficient air conditioner, Efficient electric water heater, Efficient gas/oil water heater, Solar water heater, Efficient gas cooking appliances, Efficient electric cooling appliances, Efficient lights, Efficient visual display, Efficient refrigerator, Efficient cool/hot carrier system, Fuel cell cogeneration, Photovoltaic, Building energy management system (BEMS), Efficient insulation, Eco-life navigation, Electronic newspaper/magazine etc.
Transportation	Efficient reciprocating engine vehicle, Hybrid engine vehicle, Bio-alcohol vehicle, Electric vehicle, Plug-in hybrid vehicle, Natural gas vehicle, Fuel-cell vehicle, Weight reduction of vehicle, Friction and drag reduction in vehicle, Efficient railway, Efficient ship, Efficient airplane, Intelligent traffic system (ITS), Real-time and security traffic system, Supply-chain management, Virtual communication system etc.
Industrial	Efficient technologies for boiler, industrial furnace, Independent power plant (IPP), coke oven, and other innovations such as Eco-cement, Fluidized catalytic cracking of naphtha, Methane coupling, Gasification of black liquid.
Energy transformation	Efficient coal-fired generation (IGCC, A-PFBC, Co-combustion with biomass etc), Efficient gas-fired generation, Efficient biomass-fired generation, Wind generation (onshore, offshore), Nuclear power generation, Hydro power generation, By-product hydrogen, Natural gas reforming hydrogen production, Biomass reforming hydrogen production, Electrolysis hydrogen production, Hydrogen station, Hydrogen pipeline, Hydrogen tanker, CCS (Carbon capture and storage), etc.

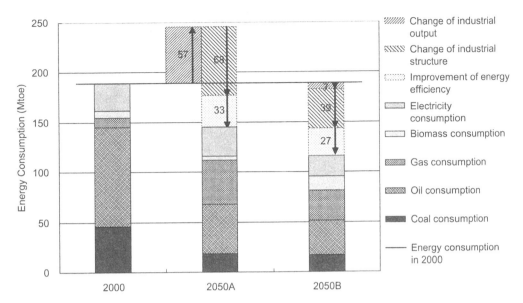

FIGURE 4 Energy demand reduction in the industrial sector.

Change of industrial output: changes in total production of primary and secondary industry.
Change of industrial structure: structural changes in primary and secondary sectors.
Improvement of energy efficiency: efficient energy devices such as furnaces and motors.

Energy demand in manufacturing processes can be classified as direct heating, steam, mechanical power, chemical reduction, refining, and others. Energy efficiencies of the technologies to fulfil those energy demands, such as furnaces, boilers and motors, have great potential for improvement.

In Scenario A, in which 2% of annual GDP per capita growth rate is assumed, the production of primary industries and secondary industries will increase. Nevertheless, it is estimated that 20% of energy savings can be achieved through the reduced output of energy-intensive raw materials and energy efficiency improvements (their energy saving effects are 68 Mtoe and 33 Mtoe, respectively).

In Scenario B, it is estimated that a further 40% of energy reduction is possible by industrial transformations and energy efficiency improvements (see Figure 4).

Therefore, 20–40% reductions in energy demand are possible through structural transformation and energy-efficient technologies.

4.1.2. Possible countermeasures to achieve LCS in the passenger transportation sector (Figure 5)

Passenger transportation demand reduces due to several factors, which include decrease in population, reduction in average trip distance via development of the 'compact city' aimed at a secure and safe society, promotion of public transformation in order to achieve urban development dedicated to vulnerable road users, and other measures.

Combining the demand reduction measures together with more efficient vehicles, such as hybrid vehicles or electric-powered vehicles, and changes in fuel towards lower carbon intensity (electricity, hydrogen and biomass) can lead to an 80% reduction in energy demand from the passenger transportation sector.

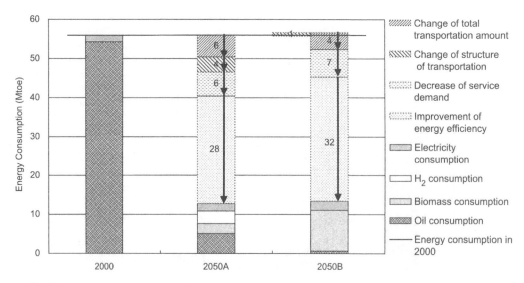

FIGURE 5 Energy demand reduction in the passenger transportation sector.

Change of total transportation volume: decrease in total transportation demand associated with population decrease.
Change of transportation structure: modal shift towards public transportation due to change of population allocation (LRT, public buses, etc.).
Decrease of service demand: shortened average trip distances associated with changes in urban structure such as compact city.
Improvement of energy efficiency: Improvement of energy efficiency of vehicles (hybrid vehicle0lightening of car body, etc.).

In Scenario A, households, offices and shopping centres are located more intensively in the urban areas as compared with Scenario B. Therefore, the average trip distance is shorter and the modal share of public transportation is higher. The energy-saving effects of intensive land use and modal shift are 6 Mtoe and 4 Mtoe, respectively. In addition, promotion of increasing density of urban structures could further reduce 6–7 Mtoe of energy consumption for both scenarios.

Note, however, that car transportation would still have the highest share in the passenger transportation sector in 2050. Technological vehicle innovations, including market penetration of electric or hydrogen vehicles, lightness of car body weight, reduced air resistance, installation of hybrid engines, and other similar measures could save 28 Mtoe of energy demand for Scenario A and 32 Mtoe for Scenario B.

In summary, 80% energy demand reductions are possible through appropriate land use and energy-efficient technologies.

4.1.3. Possible countermeasures to achieve LCS in the freight transportation sector (Figure 6)

It is possible to save approximately 70% and 60% of energy demand in Scenarios A and B, respectively, by developing rational logistics systems through information and communications technologies (ICT) and by promoting more efficient transportation vehicles.

SCM (supply chain management), which includes the appropriate management of product flows and efficient route searching through ICT, would be expected to enhance the load efficiency and lessen the volumes of returned or disposed products. In addition, advanced management of logistics networks would make the connections between small-lot cargoes by truck and large-lot cargoes by ship/train smoother, thus facilitating a modal shift. Those effects are estimated to be 3 Mtoe for Scenario A and 2 Mtoe for Scenario B.

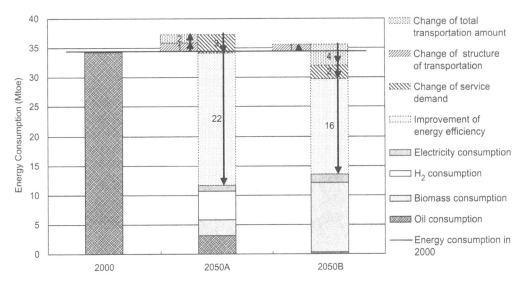

FIGURE 6 Energy demand reduction in the freight transportation sector.

Change of total transportation volume: changes in freight transportation demand associated with industrial production.
Change of transportation structure: changes in mode of transportation through modal shift.
Change of service demand: changes associated with application of rational logistic systems such as SCM (supply chain management).
Improvement of energy efficiency: improvement of energy efficiency of vehicles.

Through drastic improvements in energy-efficient vehicles in the freight transportation sector, for Scenario A, 22 Mtoe of energy would be saved with the penetration of electric or hydrogen-fuelled vehicles. In Scenario B, on the other hand, it is assumed that 50% more efficient vehicles fuelled with biomass would reduce the energy demand by 16 Mtoe in this sector.

In summary, 60–70% energy demand reductions are possible through efficient transportation management systems and energy-efficient technologies.

4.1.4. Possible countermeasures to achieve LCS in the residential sector (Figure 7)

The average lifetime of dwellings in Japan is around 35 years and a significant number of existing dwellings will be rebuilt by 2050. Therefore it is possible to create a dwelling stock that can strike a balance between comfortable living space and energy saving by a programme for the future replacement of existing buildings with highly insulated, energy-saving dwellings.

The decrease in the number of households compensates for the increase in energy demand with the rise of service demand. The reason why the service demand of Scenario A, in which people seek convenience, is almost the same as that of Scenario B, in which people prefer more sustainable ways of living, is that the energy service demand of Scenario A will decrease due to eating out, an increase in collective houses, and other new practices.

Rebuilding with highly insulated dwellings could reduce demand by about 10 Mtoe. In addition to this, technological innovations such as efficiency improvement of heat pumps (for air conditioners and electric water heaters), cooking stoves, lighting, and standby power could reduce 50% of total energy demand below the 2000 level.

CO_2 emissions from the residential sector could be almost reduced to zero by increasing the use of hydrogen and electricity which does not emit CO_2 in Scenario A, and by increasing the use of distributed renewable energy such as solar heat, solar power and biomass in Scenario B.

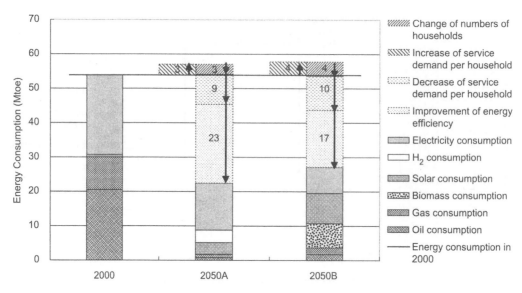

FIGURE 7 Energy demand reduction in the residential sector.

Change of the number of households: the number of households decrease both in Scenario A and Scenario B from present to 2050.
Change of service demand per household: convenient lifestyle increases service demand per household.
Change of energy demand per household: high insulated dwellings, thermo bathtub and home energy management system (HEMS) decrease service demand per household.
Improvement of energy efficiency: energy efficiency improvement of air conditioner, water heater, cooking stove, lighting and standby power.

Future developments in building technology, leading to the replacement of existing buildings with new, highly insulated dwellings, which strike a balance between comfortable living space and energy saving, reduces 50% of energy demand.

4.1.5. Possible countermeasures to achieve LCS in the commercial/service sector (Figure 8)

The demand for floor space in service industries would rise due to an increase in their outputs as the economy becomes more service-oriented in both scenarios, A and B (Figure 3). However, the labour population of the service industry would not increase much due to the overall decline in population. The number of electric appliances in offices would increase in order to improve the working environment, but insulation, BEMS and energy-efficiency improvement of appliances could reduce 40% of energy demand in 2050 (Figure 8).

The activity of hotels, restaurants and places of entertainment, which consume a lot of energy would increase with the active consumption in Scenario A. As a result, the energy consumption would increase by 12 Mtoe. However, rebuilding to high insulation standards and distributing BEMS could reduce 7 Mtoe of energy demand. In addition, high-efficiency air conditioners, water heaters and lighting could reduce 24 Mtoe of energy demand.

A combination of comfortable servicing space/working environment and energy-efficiency improvements reduces 40% of energy demand.

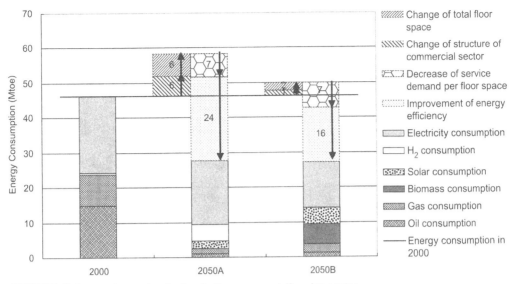

FIGURE 8 Energy demand reduction in the commercial/service sector.

Change of commercial activity: floor space increases with the rise of activity.
Change of commercial sector structure: share of energy intensive business fields such as hotels and restaurant increases.
Change of energy demand per floor space: high insulated buildings and building energy management system (BEMS) reduce energy demand.
Improvement of energy efficiency: efficient air conditioner, water heater, lighting fulfils service demand with less energy.

4.2. Low-carbon alternatives for energy supply and demand

We have examined feasible combinations of energy sources in the supply side, as shown in Figure 9, which satisfy both secondary energy demands and quantity constraints for various energy resources (Steps 4 and 5 in Figure 1). Various criteria for energy supply side include economic efficiency, uncertainty of technological innovation, public acceptance, and expert judgements in the context of the narrative scenarios. As per our estimate, required energy demands in 2050 would decrease to about 55–60% of the 2000 level owing to various kinds of innovations, even with reasonable economic growth. In addition, decarbonization of energy supply will be necessary in order to achieve the LCS. Decarbonized energy supply systems exhibit a lot of variation. In this research, it is assumed that large-scale centralized energy systems, such as nuclear power, carbon capture and storage (CCS), and hydrogen production are suitable options for Scenario A, and small-sized distributed energy systems, such as solar, wind and biomass are suitable for Scenario B.

4.3. Technology cost to achieve LCS

The future cost of countermeasures will vary depending on the direction of socio-economic development in the envisioned society. In order to achieve the envisioned society in 2050, it is necessary to actively implement industrial transformation and investment in transportation infrastructure as soon as possible. Those investments are not necessarily carried out as part of a climate change policy since they will be deployed anyway for enhancing Japan's international competitiveness, designing safe communities with comfortable levels of mobility and energy security. In this study, it is assumed that those investments, which also contribute to LCS, will occur anyway

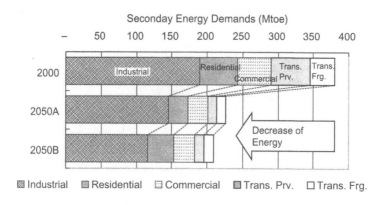

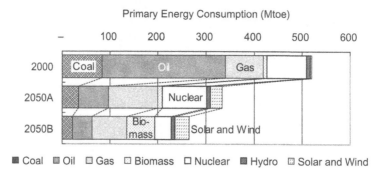

FIGURE 9 Energy demands and supply for achieving 70% reduction of CO_2 emissions (Mtoe: Million ton of equivalent).

with appropriate timing. Hence those investments are not considered in the cost analysis for achieving LCS in this study. Such investments include non-technological costs such as the cost of upgrading social and physical infrastructures, and technological and social innovations that are likely to be induced by market-based competitive forces. However, even if those investments are undertaken, it will become necessary to accelerate the diffusion of further innovative technologies to achieve LCS.

This study estimates the annual implementation cost of introducing low-carbon technologies in order to achieve a LCS targeted at 70% emission reductions in 2050. Costs of individual technologies were estimated based on a bottom-up approach. The total annual implementation cost of introducing low-carbon technologies is estimated to be 8.9–9.8 trillion yen (US$83–91 billion) for Scenario A and 6.7–7.4 trillion yen (US$62–69 billion) for Scenario B, as shown in Table 4.[1] The additional costs were estimated by subtracting the cost of existing technology from the cost of selected low-carbon technology in each sector (Figure 10). Annual additional costs in 2050 were estimated to be 1.0–1.8 trillion yen (US$9–17 billion) for Scenario A and 0.7–1.6 trillion yen (US$7–15 billion) in Scenario B. The corresponding average reduction costs are estimated in the range of ¥20,700–34,700/tC (US$190–320).

The difference between the cost of conventional technology (existing technology which is induced when no action for LCS is taken) and low-carbon technology (necessary technology to actualize 70% reduction) is considered as the additional cost.

TABLE 4 Additional cost for LCS

	Scenario A	Scenario B
Annual implementation cost (% of GDP in 2050)	8.9–9.8 trillion yen (US$83–91 billion) (0.83–0.90%)	6.7–7.4 trillion yen (US$62–69 billion) (0.96–1.06%)
Annual additional cost	1.0–1.8 trillion yen (US$9–17 billion)	0.7–1.6 trillion yen (US$7–15 billion)
Average reduction cost	24,600–33,400 yen/tC (US$230–320)	20,700–34,700 yen/tC (US$190–310)

Average reduction cost = Additional cost/Emission quantity reduced by the additional measure.

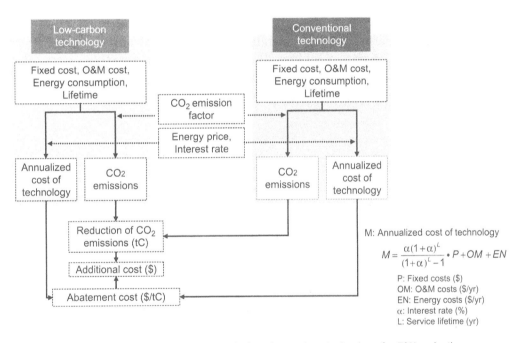

FIGURE 10 Calculation of additional cost to induce low-carbon technology for 70% reduction.

5. Conclusions

Through this research, we conclude that CO_2 emissions in 2050 can be reduced by 70% from the 2000 level, both in Scenario A and Scenario B, under an acceleration of research, development and deployment of new technologies. 2050's GDPs in Scenarios A and B are set to about double and 1.5 times the 2000 level, respectively. Expected future innovations will lead to the reductions in energy demand by 40–45% from the 2000 level while maintaining GDP growth and improving service demands. In addition, the decarbonization of energy supply boosts CO_2 reductions to 70% below 2000 level. In each scenario, strategies for realizing LCS include three key elements:

demand reduction through rationalization of energy use; development and deployment of energy-efficient technologies; and decarbonization of energy in supply side.

Effective countermeasures in Scenario A are energy-efficiency options on the demand side, such as implementation of energy-efficient appliances in the industrial, residential, commercial and transportation sectors, and fuel-switching options from conventional energy sources to low-carbon energy sources, such as nuclear power and hydrogen. In Scenario B, the use of low-carbon energy, such as biomass and solar energy, on the demand side would result in drastic reductions of CO_2 emissions.

Although CO_2 reductions by sector vary according to the scenario, both scenarios share many technology options. These options involve 'no-regret' investments, which reduce the energy costs and are profitable. Research and development activities for such technologies yield desirable outcomes for society. The technology options that take long periods of time to implement, such as hydrogen, nuclear power and renewable-based energy systems, require early, well-planned strategies that take uncertainties into consideration.

In order to achieve LCS by 2050 without missing opportunities for various investments on capital formation and technology development, it is necessary to establish the national goals (i.e. the vision of LCS, rather than target rates of reduction) at an early stage, establish the abatement schemes, and realize a society that internalizes the negative externalities of CO_2 emissions. In this process, the social and technological innovations would be accelerated, thereby imparting an advantage with regard to international competition in the future low-carbon world.

Acknowledgement

This study is sponsored by Global Research Environment Fund (S-3-1, GREF), MOEJ. We appreciate its generous funding.

Note

1. All US dollar figures in this article have been calculated assuming the rate of ¥107.39 = US$1, the approximate exchange rate prevailing on the Tokyo Foreign Exchange Market on 31 March 2005.

References

Cabinet Office, 2005, *Japan's Visions for the 21st Century*.

Fukuda, Y., 2008, *Special Address by H.E. Mr. Yasuo Fukuda, Prime Minister of Japan on the Occasion of the Annual Meeting of the World Economic Forum, Congress Center, Davos, Switzerland* [available at www.kantei.go.jp.p.fn.to/foreign/hukudaspeech/2008/01/26speech_e.html].

IPCC, 2007a, 'Summary for Policymakers', in: *Climate Change 2007: Impacts, Adaptation and Vulnerability*. Contribution of Working Group II to the Intergovernmental Panel on Climate Change, Fourth Assessment Report, Cambridge University Press, Cambridge, UK.

IPCC, 2007b, 'Summary for Policymakers', in: *Climate Change 2007: Mitigation*. Contribution of Working Group III to the Intergovernmental Panel on Climate Change, Fourth Assessment Report, Cambridge University Press, Cambridge, UK.

Matsuoka, Y., Fujino, J., Kainuma, M., 2007, *Aligning Climate Change and Sustainability: Scenarios, Modeling and Policy Analysis*, Asia–Pacific Integrated Modeling Team, Centre for Global Environmental Research (CGER), National Institute for Environmental Studies (NIES), Japan, CGER-I072-2007 [available at www-cger.nies.go.jp/publication/I072/I072.pdf].

METI, 2005, *Strategic Technology Roadmap in Energy Field–Energy Technology Vision 2100*.

■ research article

The role of international drivers on UK scenarios of a low-carbon society

NEIL STRACHAN[1]*, STEPHEN PYE[2], NICHOLAS HUGHES[1]

[1] King's College London, London, UK
[2] AEA Energy and Environment, Didcot, UK

An integrated set of low-carbon society (LCS) scenarios for the UK were analysed using the UK MARKAL Macro (M-M) model. A $100/tCO_2$ carbon price scenario was compared with long-term LCS scenarios with a domestic 80% CO_2 reduction target. As M-M is a national-level model, a set of five international drivers were investigated, and grouped under *Annex I consensus* and *Global consensus* assumption sets. For economy-wide results the inclusion of international aviation and potential large-scale purchases of CO_2 permits (when available) are most important. For sectoral implications, all international drivers considered here are important; for example in divergent overall size and configuration of the UK electricity sector. The carbon price scenario and set of 80% LCS targets scenarios give GDP losses rising from 0.36% to a range of 1.64–2.21% in 2050. This steep cost convexity in deep CO_2 reductions represents increasing efforts to decarbonize the UK energy system, and the further impact of key international drivers. This illustrative analysis demonstrates that UK policy makers should be cognisant of, and flexible with respect to, international strategies on LCS and emission reduction targets.

Keywords: carbon pricing; climate change; climate stabilization; development pathways; energy systems; low-carbon society; scenario modelling; UK

Une série intégrée de scenarios de sociétés sobres en carbone (LCS) pour le Royaume-Uni fut analysée à l'aide du modèle MARKAL Macro (M-M) du Royaume-Uni. Le scénario pour un prix du carbone de $100/T CO_2 fut comparé aux scénarios LCS à long terme à objectif de réduction domestique en CO_2 de 80%. Vu que M-M est un modèle à échelle nationale, cinq influences internationales furent analysées, et groupées selon les hypothèses de consensus dans l'Annexe 1 et consensus mondial. En ce qui concerne les résultats pour l'économie dans sa totalité, l'inclusion de l'aviation internationale et l'achat potentiel de permis de CO_2 à grande échelle (selon leur disponibilité) sont des facteurs particulièrement importants. Pour les conséquences sectorielles, toutes les influences internationales prises en compte sont importantes, par exemple vis-à-vis des différences dans la taille globale et la configuration du secteur électrique du Royaume-Uni. Le scénario du prix du carbone et la série de scénarios à objectif LCS de 80% mènent à des pertes de PIB allant de 0.36% à une fourchette de 1.64% à 2.21% en 2050. La convexité abrupte du prix pour des réductions profondes en CO_2 illustrerait les efforts accrus de décarbonation du système énergétique britannique, ainsi que l'effet supplémentaire des principales influences internationales. Cette analyse illustrative démontre que les décideurs du Royaume-Uni devraient être informés et flexibles par rapport aux stratégies internationales sur les LCS et les objectifs de réduction des émissions.

Mots clés: changement du climat; fixation du prix du carbone; modélisation de scenarios; Royaume-Uni; société sobre en carbone; stabilisation climatique; systèmes énergétiques; voies de développement

■ *Corresponding author. E-mail*: neil.strachan@kcl.ac.uk

CLIMATE POLICY 8 (2008) S125–S139

doi:10.3763/cpol.2007.0489 © 2008 Earthscan ISSN: 1469-3062 (print). 1752-7457 (online) www.climatepolicy.com

1. Introduction

Under the auspices of the G8 International Dialogue on Climate Change,[1] the UK and Japanese governments set up a technical project on achieving a low-carbon society (LCS), with the goal of investigating long-term (to 2050) LCS pathways through technological solutions, innovation and behavioural change. As part of this project, a range of international teams (with a strong developing-country participation), undertook modelling of long-term carbon emission targets, with a core focus on the linkages between global and national LCS drivers.

This article describes the analysis of LCS scenarios using the UK MARKAL-Macro (M-M) model, a hybrid technology optimization model linked to a neoclassical growth model (see Section 2). This model was used for underpinning analysis on long-term carbon reductions in the 2007 UK Energy White Paper (DTI, 2007). However, as this is a national model, the role of international drivers is crucial for the quantification of the costs and pathways of a future LCS. Section 3 describes a set of key international drivers and how these are quantified within the UK-M-M analysis. Section 4 details the implementation of a set of scenarios run for this article, based on the common runs for all modelling papers under this LCS project:

- *Baseline*: where UK energy use and CO_2 emissions occur in the absence of new climate policy
- *Carbon price*: where a carbon tax (in US$ per tonnes of carbon dioxide (tCO_2)) is applied, starting at $10/$tCO_2$ in 2020 and rising exponentially to $100/$tCO_2$ in 2050
- *Carbon-plus*: under a framework of a global target of a 50% reduction in CO_2 emissions, the UK leads on international mitigation efforts and achieves a domestic CO_2 reduction of 80% by 2050
- Sensitivity runs on *Carbon-plus*, based on the role of international drivers (see Section 3).

Section 5 details the results from this modelling exercise. Section 6 discusses UK policy implications for long-term LCS in the context of global actions and resultant international drivers.

2. Overview of the UK M-M model

MARKAL (acronym for MARKet ALlocation) is a widely applied, dynamic, technology-rich linear programming (LP) optimization model (see Loulou et al., 2004). As a partial equilibrium energy systems model, its objective function minimizes total (discounted) capital, fuel and operating costs for resource, process, infrastructure, conversion and end-use technologies. The model represents each element of the integrated energy system with time-varying technical and economic parameters. The connection and interaction of the nodes form the model topography and hence the energy system pathways. A wide-ranging application of constraints to represent policy and physical constraints, implementation of all taxes and subsidies, and inclusion of base-year capital stocks and flows of energy, enable the calibration of a model to a particular local, national or international (multi-region) energy system.

A major methodological extension to MARKAL has been the integration of this rich technological characterization of an energy system with a macroeconomic neoclassical growth model (Manne and Wene, 1992). This general equilibrium MARKAL-Macro (M-M) model now has its objective function as the maximization of the discounted log of utility (consumption), within which energy system costs are minimized. The output (production) of the economy via the Macro module is used for consumption, investment and energy costs in a single economic sector with perfect foresight.

In summary (explained in detail in Manne and Wene, 1992), M-M has four major features:

- An aggregated demand feedback from changes in energy prices.
- Autonomous demand changes to allow scenario analysis where some energy demands are decoupled from economic growth.
- An explicit calculation of gross domestic product (GDP) and other macro variables (consumption, investment).
- Energy systems effects within MARKAL (i.e. changes in technology mixes and competition for fuel and infrastructures in long-term energy system evolution).

The construction of the UK M-M model entails the first UK application of the general equilibrium Macro linkage as well as definition of the specific characteristics of the entire UK energy system via MARKAL. A full description is given in Strachan et al. (2007). Key inputs into the model include:

- base levels for global resource and national supply curves
- detailed energy service demands in units of useful energy
- future energy technologies (based on future vintages or learning rates)
- economic parameters (e.g. discount rates, price elasticities).

A complete description of all input parameters, including the sources and rationale for assumptions on the resources, technology costs and classifications, energy service demand derivation and global model parameters is given in the model documentation (Kannan et al., 2007).

3. Enabling international drivers in UK M-M model

A core strength of a national energy model such as UK M-M is its detailed depiction of physical, market and policy aspects of a country-level energy system. However, single-region models must recognize key international drivers that will have economic and technological impacts. These drivers include:

- technology costs
- fossil fuel resource prices
- supply of imported resources, including biomass
- international aviation emissions
- trading mechanisms for international CO_2 emission reductions permits.

Conventionally, these drivers are treated as exogenous assumptions, which is adequate when analysing localized national energy policies in isolation (e.g. Contaldi et al., 2007) but less than adequate when considering long-run carbon emissions reductions as part of a regional or global mitigation strategy. An alternative approach is to utilize global models broken into key regions (e.g. Weyant, 2004), but this approach typically loses country-level detail on the UK energy system (as discussed in Section 2).

As part of this article, scenario analysis is undertaken on these key international drivers to investigate the sensitivity of the UK model under long-term deep carbon reductions (our *Carbon-plus* case). These drivers are undertaken under two storylines.

The first storyline, known as *Annex I consensus* in this article, describes a situation where Annex I countries under the Kyoto Protocol (including the USA) move towards significant carbon emissions reductions of at least 60% by 2050, relative to 1990 levels. They reflect the UK initiative in implementing binding carbon budget legislation (DEFRA, 2007). The UK's long-term carbon emissions reduction is at least 80% by 2050. This storyline assumes that other Annex I countries will also pursue complementary goals, while non-Annex I countries do not, but continue on a business-as-usual pathway.

The second storyline, known as *Global consensus* in this article, describes a situation where all countries contribute towards a global target of an absolute reduction in global emissions of around 50% by 2050 (G8 Communiqué, 2007). Recognizing the many uncertainties related to the roles of non-CO_2 GHGs, carbon sinks, and so on, this CO_2 emission pathway is consistent with an atmospheric stabilization target of 550 ppm. Under this scenario, the UK achieves at least an 80% reduction, Annex I countries achieve at least a 60% reduction, with non-Annex I countries achieving at least a 30% reduction under a future long-term international mechanism.

Taking each international driver in turn, we outline our assumptions for each under the two storylines.

3.1. Technology costs (vintages and learning curves)

A key energy model input is the future representation of technology costs, based on the evolution of technical performance and capital costs. In UK M-M, fossil extraction, energy processes (e.g. refineries), infrastructures, transport, buildings, industrial and many electricity technologies utilize vintages to present improvements through time. Less mature renewable and zero-carbon electricity,[2] micro-generation, high-efficiency end-use appliances and hydrogen transport technologies have exogenously calculated future costs, which are based on both expected learning rates (through R&D and learning-by-doing) and the total uptake of these technologies.

Within a global economy, the UK is a price taker across most emerging energy technologies. Future cost reductions in technologies will thus be a function of the global investment levels in R&D and experience of technology implementation. Global technology uptake is based on projections from the World Energy Technology Outlook (European Commission, 2005), whilst assumptions on learning rates are based on McDonald and Schrattenholzer (2002).

Under the scenario *Global consensus*, rates of learning accelerate in future years, due to the increased penetration and hence experience of technology implementation driven by global mitigation requirements. We assume that the full potential for learning, as outlined in the above sources, is realized under this scenario. Under the *Annex I consensus* scenario, only half of the potential learning is assumed. This is an indicative and parametric assessment, based on projections (e.g. EIA, 2007) that developing countries will constitute a majority of energy use and hence technology deployment; however, the source of learning – although much harder to measure – is more likely to remain dominated by developed countries.

3.2. Imported fossil resource supply curves

The UK M-M model utilizes resource costs based on exogenous baseline price assumptions from standard UK government forecasts (DTI, 2006). As the model then depicts domestic and imported fossil resources using supply curves, multipliers calibrated from relative prices are used to translate these into prices both for higher-priced supply steps as well as for imported refined fuels.

TABLE 1 Exogenous base fossil fuel import prices (various units, 2000 prices)

Year	Baseline			High prices			Low prices		
	Oil $/bbl	Gas p/therm	Coal $/GJ	Oil $/bbl	Gas p/therm	Coal $/GJ	Oil $/bbl	Gas p/therm	Coal $/GJ
2010	40.0	33.5	1.9	67.0	49.9	2.4	20.0	18.0	1.4
2020	45.0	36.5	1.8	72.0	53.0	2.6	20.0	21.0	1.0
2030	50.0	39.6	2.0	82.0	59.0	2.8	25.0	24.0	1.2
2040	55.0	42.6	2.2	82.0	59.0	3.0	30.0	27.0	1.3
2050	55.0	42.6	2.2	82.0	59.0	3.0	35.0	30.0	1.5

Source: DTI (2006).

In this article, the baseline prices represent the *Annex I consensus* storyline, where future demand in non-Annex I countries continues to increase and push prices higher. Under a *Global consensus*, we assume that, due to mitigation efforts across all world regions, demand for fossil fuels reduces because of lower levels of energy consumption or switching to lower carbon fuels. As a result, the price of these resources is relatively lower in future years than assumed in the baseline. For this storyline, we have assumed the 'low prices' in Table 1.

3.3. Supply of imported biomass resources

Previous analyses (Strachan et al., 2007) have shown that under a severely carbon-constrained system, biomass plays an important role in future years, particularly in the UK transport sector. The amount of biomass that the UK can use is constrained by both domestic production capacity and the level of imports. Constraints on UK imports of biomass are determined by the UK's share of total GDP applied to estimates of global supply (detailed in (Strachan et al., 2007), and applied to the different resource steps. This model baseline is considered to be representative of the *Annex I consensus* scenario.

With increasing international use of biomass expected under the *Global consensus* scenario, we introduced a constraint on the UK's share of that global supply of biomass of 1.1 EJ per annum in 2050, growing linearly from zero in 2000. This was based both on the emphasis on sustainable biomass production, as well as a shift in the UK's share of imports from a GDP basis to an equal per capita use basis. WWF International (2007) estimated that annual sustainable biomass with an energy content of between 110 and 250 exajoules (EJ) is feasible by 2050, at a global level.

3.4. Inclusion of the international aviation sector

In current inventory methodologies used for reporting under the UNFCCC,[3] carbon emissions from international aviation are not counted in domestic budgets. Under the *Annex I consensus* scenario, this approach continues, with the assumption that Annex I countries will not include international aviation in carbon budgets due to competitiveness and political considerations.

With a focus on significant levels of mitigation across all global regions in the *Global consensus* scenario, international aviation emissions are assumed to be accounted for within domestic carbon inventories. As a rapidly growing sector, the inclusion of international aviation is an important driver, as it will impact on the UK's ability to make significant future carbon reductions. Air travel also entails secondary radiative forcing effects due to high altitude emissions, and this is reflected through a standard multiplier of 2.5 (DfT, 2004). In this analysis we have assumed total aviation energy use is held at 2010 levels,[4] representing UK airport expansion restrictions.

3.5. Emissions purchases, and development of a global MACC curve

For the UK to meet future carbon reduction targets, the purchase of emission credits, resulting from mitigation overseas, could be critical. To model the potential for the UK to account for international action domestically, we have introduced a set of long-run regional marginal abatement cost curves (MACCs) into the model (for a detailed description, see Sands, 2004).

Regional long-run marginal cost curves[5] developed in the SGM model were used to represent the potential cost of purchasing international emission credits. Cost curves for 13 world regions ranged between \$3 and \$210/tCO$_2$ – and had associated carbon reductions at each price step.[6] Based on baseline projections for energy use and carbon emissions (EIA, 2007), it was assumed that these regions had their own domestic targets (through 2050).

In the current UK Government consultation on the draft Climate Change Bill (DEFRA, 2007), the use of international emission credits is still uncertain and somewhat controversial. The Bill proposed that

> ... emissions reduction achieved overseas but paid for by UK entities is to be counted towards the targets and budgets. This does not mean that all (or an unlimited amount of) emissions reduction effort should or would be achieved overseas. Guidance on the degree to which emissions reductions should be achieved domestically are contained in the international principle of 'supplementarity'.[7]

Recognizing both this principle and guided by the current UK policy debate, in the sensitivity runs below the UK is restricted to a ceiling of 50% of domestic targets to be met by permits purchased from overseas (through trading or other mechanisms), if it is cost-effective to do so.

The following assumptions were made:

- Continuing the theme of supplementarity, each region's domestic abatement requirements would use the cheapest mitigation measures first,[8] resulting in traded permits being more costly.
- The UK continues to be in the vanguard of emission mitigation efforts with its 80% reduction target (in the *Carbon-plus* cases) resulting in its potential purchases of these more expensive emissions credits.
- MACCs only include energy CO$_2$ mitigation measures, not non-CO$_2$ GHGs or land-use change. Inclusion of other GHG abatement options would be likely to increase supply and reduce costs.
- MACCs are based on current projections of technology development, although it is noted that more optimistic assumptions could lead to greater cost-effective potential (Sands, 2004).
- Regional MACCs are robust to technology and energy price changes in other regions (Ellerman and Decaux, 1998), although later studies illustrate that this is not necessarily the case (Klepper and Peterson, 2006).
- The UK was considered separately, although implicitly contained within the global MACCs. This approximation is reasonable, as the UK represents a global GDP share of only 3.5%, projected to fall to 2.5% in 2050 (EIA, 2007).

TABLE 2 Available international permits ($MtCO_2$) under marginal abatement costs curves (MACCs)

$MtCO_2$		2030	2050						
UK constraint (50% supplementarity)		128	242						
UK reduction requirement		260	484						
($/tCO_2$)		3	8	14	26	52	104	210	Total
Annex 1 consensus MACC	2030	0	0	0	0	27	95	44	166
	2050	0	0	0	0	0	0	0	0
Global consensus MACC	2030	0	0	0	0	13	54	41	108
	2050	0	0	0	0	0	0	0	0
Global consensus sensitivity MACC	2030	85	62	94	97	125	193	142	799
	2050	85	62	94	97	98	98	98	634

Three MACC curves were developed (as used in the model scenarios in Section 4):

1. *Annex I consensus MACC*. Annex I mitigation and supply, with permits only available for purchase from Annex I countries, who move towards a 60% CO_2 reduction target. The UK (with a higher 80% reduction target) can access up to 5.3% of this Annex I market, based on a proportion of UK vs. Annex I GDP.
2. *Global consensus MACC*. Global mitigation and supply, with all regions selling permits following Annex I attaining a 60% reduction and non-Annex I attaining a 30% absolute reduction in CO_2 emissions (LCS-CPM1). The UK can access up to 3.5% of the market (falling to 2.5% by 2050), based on a proportion of UK vs. global GDP.
3. *Global consensus sensitivity MACC*. Annex I mitigation and supply plus non-Annex I sales assuming no developing-country emissions targets. This results in higher levels of supply and cheaper prices. This MACC is used in sensitivity run LCS-CPM2. The UK can access up to 3.5% of the market, based on a proportion of UK vs. global GDP, falling to 2.5% in 2050.

Table 2 summarizes (data for 2030 and 2050) what the UK is permitted to use versus the availability of permits for the UK in future years.

4. UK M-M model runs of LCS scenarios

Four core model runs are shaded in grey in Table 3.

- *Baseline*. Under the UK M-M baseline, the energy system evolves without any new carbon policy but reflecting legislated UK Government policies.[9] The baseline is identical to that in the 2007 Energy White Paper (see DTI, 2007, for a full listing of included policies) and is labelled LCS-BAS.[10]
- *Carbon price* of $10 in 2010 exponentially rising to $100 in 2050.[11] The introduction of this carbon tax on the UK energy system approximates to a 55% reduction in carbon emissions by 2050. This run is labelled LCS-CT.

TABLE 3 UK M-M model runs to assess *Annex 1 consensus/Global consensus* scenarios

Model run	Run label	Run description
Baseline	LCS-BAS	UK energy system evolves without any new carbon policy but reflecting legislated UK Government policies
Carbon price	LCS-CT	$10 in 2010 exponentially rising to $100 in 2050
Carbon-plus: Annex 1 consensus	LCS-CP	UK energy system moves to 80% carbon reductions by 2050. International driver assumptions reflect *Annex I consensus*
Carbon-plus with international aviation	LCS-CPAV	80% reduction in 2050 relative to 1990 levels with international aviation sector included
Carbon-plus with enhanced technological learning	LCS-CPLN	Rates of technological learning increase through learning-by-doing across all global regions
Carbon-plus with lower fossil resource prices	LCS-CPLP	Reduction in fossil fuel demand globally due to mitigation targets leading to lower resource prices
Carbon-plus with biomass import restrictions	LCS-CPBM	Reduction in the availability of biomass imports due to supply country's sustainable production and domestic mitigation
Carbon-plus with global MACC	LCS-CPM1	Global marginal abatement cost curve assuming all regions involved in carbon permit market as buyers and suppliers
Carbon-plus with global sensitivity MACC	LCS-CPM2	Global marginal abatement cost curve assuming Annex I as buyers and suppliers, and non-Annex I as sellers only
Carbon-plus: Global consensus	LCS-CPGC	CPAV, CPLN, CPLP, CPBM, CPM2 assumptions used in sensitivity runs combined to produce a *Global consensus* run

▦ *Carbon-plus: Annex I consensus* and *Global consensus*. These two integrated model runs consider the economic and technological implications for the UK energy system of moving to more stringent carbon reductions of 80% by 2050 relative to 1990 emission levels. In the context of the drivers described earlier, the first model run reflects *Annex I consensus*, where only OECD countries move towards a long-term LCS goal. The second model run reflects *Global consensus*, where all regions move towards a long-term LCS. Note that Section 3 summarizes these international driver combinations for the storylines for *Annex I consensus* and *Global consensus*.

These core runs were supplemented with sensitivity runs based on alternative international drivers to explore the relative impact of each.

5. Results

The results of the model runs provide insights into how the UK energy system could respond to international drivers on scenarios of long-run deep carbon constraints. This section focuses on the implications of the scenario runs for system evolution, technology pathways and economic impacts.

5.1. System evolution

Figure 1 details final energy demand from years 2000 through 2050. Under the *Base case* (LCS-BAS), total energy demand decreases to 2030 due to the uptake of more efficient technologies, particularly in the transport sector, before rising through 2050 due to continued growth in overall energy service demand levels. The introduction of a carbon tax (LCS-CT) and an 80% CO_2 reduction target under the *Annex I consensus* (LCS-CP) lead to increasingly lower levels of final energy demand, through augmented take-up of upstream and end-use efficient technologies, and through behavioural change leading to declines in energy service demands through an aggregated price elasticity.

The inclusion of the aviation sector (LCS-CPAV) under an 80% reduction leads to even more significant reductions in demand, while the other sensitivities show smaller changes based on trade-offs between efficiency and fuel-switching to meet the same carbon constraint. Similarly, access to international MACCs introduces another trade-off between domestic efficiencies and purchased emissions reductions. This is particularly true in the sensitivity case where Annex I countries are the sole purchasers (LCS-CPM2), which allows the UK to remain on a much higher energy-use trajectory.

The combined *Global consensus* case (LCS-CPGC) is primarily influenced by the inclusion of the aviation sector; final energy levels, however, do not decline as quickly in 2020 and 2030 due to the availability of carbon permits, as consumption levels are not constrained to the same extent.

Focusing on a sectoral level, Table 4 compares CO_2 emissions by sector in 2050. Under tightening carbon constraints (to LCS-CT and then to LCS-CP), the electricity sector is the first to decarbonize (despite growth in electricity generation). This increase in zero-carbon electricity assists end-use sectors to decarbonize, as for example increasing the amounts of residential heat that are provided via electricity. Transport is the last sector to decarbonize but still makes a significant contribution by 2050, aided by the use of zero-carbon hydrogen production (all of which flows to the transport sector).

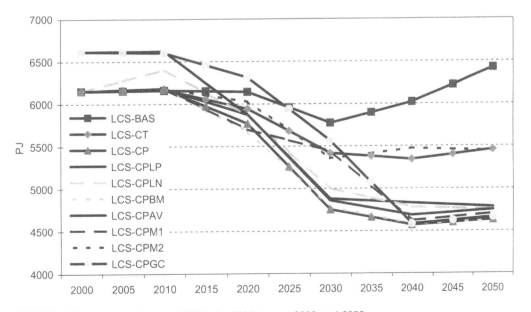

FIGURE 1 Final energy demand (PJ) in the UK between 2000 and 2050.

TABLE 4 Sectoral CO_2 emissions share in 2050

2050	[2000]	LCS-BAS	LCS-CT	LCS-CP	LCS-CPLP	LCS-CPLN	LCS-CPBM	LCS-CPAV	LCS-CPM1	LCS-CPM2	LCS-CPGC
Upstream	4.5%	3.8%	7.7%	12.5%	14.3%	12.9%	13.0%	12.9%	12.7%	5.9%	12.6%
Agriculture	0.4%	0.6%	1.2%	2.4%	2.5%	2.4%	2.4%	2.2%	2.4%	0.9%	2.2%
Electricity	35.0%	42.2%	11.7%	15.7%	6.1%	13.2%	10.1%	8.1%	13.4%	25.2%	4.6%
Industry	13.3%	11.1%	13.0%	17.2%	18.0%	17.4%	17.0%	13.9%	17.3%	10.6%	13.6%
Residential	0.0%	7.5%	5.1%	5.5%	0.3%	2.8%	10.1%	0.7%	8.1%	11.6%	0.0%
Services	16.4%	13.0%	23.6%	19.1%	22.2%	28.7%	18.5%	0.7%	21.1%	17.3%	0.3%
Hydrogen	4.9%	4.4%	3.3%	11.8%	14.0%	13.5%	12.7%	12.3%	13.5%	2.7%	12.1%
Transport	25.5%	17.5%	34.4%	15.8%	22.6%	9.1%	16.0%	49.1%	11.6%	25.8%	54.6%
Total (MtCO$_2$)	540.3	599.2	268.0	115.4	110.5	115.0	114.3	123.6	115.1	357.1	123.6

International drivers, with the exception of international aviation and MACC runs, have modest impacts on relative sector abatement levels. The inclusion of international aviation means that larger reductions are required by 2050, including additional reductions in the residential and service sectors on an end-use sector basis. Under the CPM2 sensitivity, with Annex I as major buyers of permits, the pressure for domestic reductions is very considerably eased. In the *Global consensus* scenario, no permits are available by 2050, so domestic emission reductions are consistent with the international aviation case.[12] In all scenarios, aviation, upstream emissions[13] and some industrial sub-sectors retain the majority of the remaining domestic emissions budget.

5.2. Technology pathways
The electricity sector is a key sector under LCS scenarios. A year 2050 comparison by generation fuel is detailed in Figure 2. In 2000, the UK generation sector was dominated by coal (31%), gas (37%) and nuclear (22%). Under the base case (LCS-BAS), the system evolves by 2050 to be dominated by next-generation coal (80% of generation) supplemented by modest growth in renewable technologies. Coal is preferred in the absence of any carbon constraints to other forms of generation such as gas, due to the lower resource costs and constraints on the UK natural gas network. The introduction of an escalating $100 carbon tax results in less electricity production and significant changes to the generation mix, with low-carbon technologies dominating – coal CCS accounts for 39%, with contributions from biomass (18%) and wind (13%).

Under stringent 80% carbon reductions, the power sector grows substantially as end-use sectors are forced to switch to zero-carbon electricity as a primary energy carrier. By 2050 the LCS-CP *Annex I consensus* sees the key technologies of coal CCS, nuclear and wind each contributing between 25% and 30% of generation output (see Figure 2). International drivers have a significant impact on this sector, even though (as discussed later) aggregate economic metrics show much smaller impacts. This is probably due to the closeness in levelized costs of many of the technology choices, and the subsequent sensitivity to changes in drivers.

Under a lower fossil resource price case with regard to gas (LCS-CPLP), natural gas CCS becomes the main source of generation due to gas becoming more cost-competitive.[14] Under the accelerated

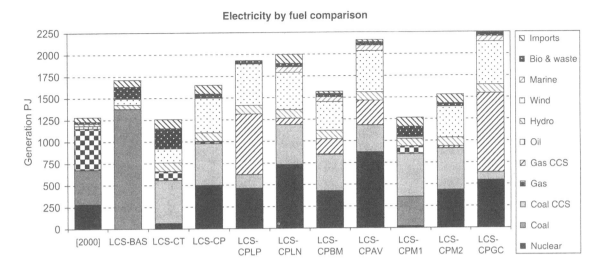

FIGURE 2 Electricity generation (PJ) by type in the UK in 2050.

learning case (LCS-CPLN), the relative cost advantage for gas CCS becomes even more important, accounting for 56% of total generation in 2050. With the inclusion of international aviation (LCS-CPAV), much higher levels of generation are observed. This appears to be because additional effort is required to abate emissions due to the inclusion of this sector; more zero-carbon electricity is therefore used in buildings end-use sectors. The ability to purchase international emission permits considerably eases the requirement for zero-carbon electricity generation, notably in the retention of conventional coal plant (LCS-CPM2). Finally, in the (LCS-CPGC) *Global consensus* case, nuclear, gas and coal CCS and wind generation continue to be the main generation options although now supplemented by advanced renewable technologies, notably marine.

The model exhibits a range of trade-offs in efforts to decarbonize among sectors under alternative scenarios. Another core sector is transport (illustrated in Table 5). By 2050, without any carbon constraints, liquid fossil fuels are still dominant (although with a greater use of hybrid vehicles). Hydrogen is also used in specific modes (bus, HGV) accounting for 13% of transport energy use. Due to the move to more efficient vehicles and hydrogen under the base case, introducing a carbon tax has a limited impact. In 2050, lower levels of petrol consumption are observed, primarily due to an increased take-up of hybrid vehicles.

Under the 80% *Carbon-plus* scenario, more radical changes are now observed in the transport sector, with diesel replaced by biofuels, in particular second-generation biodiesel. Methanol also becomes an important replacement for petrol. Some sensitivity cases diverge significantly from the *Carbon-plus* run in 2050, including advanced learning where hydrogen technologies penetrate additional modes, and the international aviation cases where increased use of jet fuel highlights the lack of cost-effective options to decarbonize this growing transport sector. Other sensitivity cases do not diverge, notably the restricted biomass case, although this limit does ensure that the use of imported biomass is overwhelmingly used by the transport sector, hence reducing the freedom to use biomass in the power sector.

Finally, in addition to technology pathways, fuel-switching and efficiency gains in key sectors as discussed above, the model trades off demand reductions, reflecting an aggregated consumer behaviour response to changes in price. By 2050, the LCS-CT case realizes demand reductions of 8–14%, which is increased to 15–24% in the LCS-CP cases due to the higher CO_2 shadow price

TABLE 5 2050 comparison of transport fuels (PJ)

2050	LCS-BAS	LCS-CT	LCS-CP	LCS-CPLP	LCS-CPLN	LCS-CPBM	LCS-CPAV	LCS-CPM1	LCS-CPM2	LCS-CPGC
Biodiesel	33.2	66.5	65.0	64.6	66.5	64.5	64.5	66.5	56.7	61.3
2nd generation biodiesel	0.0	0.0	656.8	653.1	672.8	652.5	652.6	672.0	0.0	620.1
Diesel	737.3	702.7	24.7	24.6	24.8	24.3	23.8	24.6	699.5	23.7
Electricity	71.1	76.6	65.2	64.2	64.2	67.4	65.0	65.5	75.5	62.1
Ethanol	37.5	29.0	100.0	15.6	2.4	100.0	100.0	223.5	29.1	12.5
Hydrogen	223.2	201.8	193.9	196.8	398.3	194.2	192.1	175.3	218.9	197.8
Jet fuel	96.9	89.1	77.6	77.7	77.9	77.0	246.1	77.6	88.2	240.5
Methanol	0.0	0.0	142.0	74.2	0.0	142.0	142.0	142.0	0.0	142.0
Petrol	539.7	417.0	137.1	225.0	34.7	138.9	135.1	73.5	419.2	180.2

(and hence higher unit energy costs). Generally an end-use demand with less technological options sees a greater behavioural response. The impact of the various international sensitivities (notably international carbon permits and the inclusion of international aviation) leads to modest changes via lower and higher demand responses as energy unit prices decrease and increase, respectively.

5.3. Economic impacts

Marginal CO_2 costs (as a result of the imposition of an overall carbon constraint) are detailed in Figure 3, and provide some insights into the impact of different international drivers on the costs of abatement. Marginal costs under the *Carbon-plus (Annex I consensus)* case are much higher than the $100/tCO_2$ *Carbon tax* case, and rise to $402/tCO_2$ by 2050 (or $1,470/tC). The impact of the unrestricted MACC in LCS-CPM2 is considerable, reducing marginal costs to a similar level as observed under the *Carbon tax* case.

Including international aviation increases marginal costs to around $490/tCO_2$ by 2050 (or $1,790/tC) due to the additional abatement action required. Other international sensitivities have smaller effects on the margin. The *Global consensus* scenario has lower marginal costs in 2020 and 2030 due to the opportunity to purchase permits, but costs rise sharply by 2040 as this opportunity no longer exists, and other changes to international drivers take cumulative effect. By 2050, marginal costs have risen to $589/tCO_2$.

Base case GDP rises (calculated using £ at the value of the year 2000) from around £1 trillion in 2000 to £2.8 trillion in 2050 with a much more modest growth in energy systems costs (the energy sector falls from around 9% to 5.5% of GDP).[15] However, the imposition of a stringent 80% CO_2 reduction scenario still entails noticeable percentage GDP losses.

Figure 4 details percentage GDP losses relative to the base case for all constrained scenarios. In 2050, the *Carbon tax* case shows GDP losses of only 0.33%, but this rises considerably to 1.64% (or $83 billion) in the *Carbon-plus (Annex I consensus)* case, indicating the difficulty of more stringent decarbonization for the UK. The ordering of GDP losses for the sensitivity cases based on international

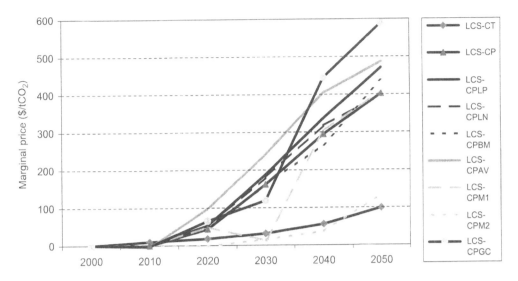

FIGURE 3 UK marginal CO_2 costs, 2000–2050, in $(2000)/tCO_2$.

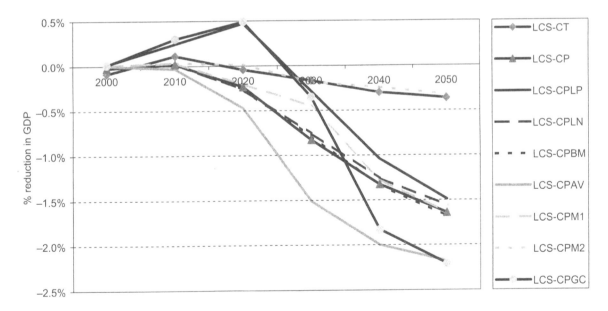

FIGURE 4 Percentage reductions in GDP relative to the base case, 2000–2050.

drivers is similar to the marginal CO_2 price results, and gives a central GDP loss range of 1.49–1.68%. Adding international aviation results in much larger impacts and this is reflected in the *Global consensus* scenario with a GDP loss in 2050 of 2.21%. Scenarios with access to lower fossil fuel prices facilitate increased economic growth in 2010–2030 before the emission constraint tightens. Similarly the availability of international permit purchases results in significantly reduced GDP costs.

6. Discussion and policy implications

This article describes analysis of low-carbon society (LCS) scenarios using the UK MARKAL-Macro (M-M) hybrid model. This modelling approach delivers optimal energy system (supply-side and demand-side) configurations assuming competitive markets and perfect foresight. As such, long-term target setting by government is a critical component, especially when analysing very deep UK CO_2 reductions. Long-term LCS are analysed via a *Carbon price* scenario (rising to $100/$tCO_2$ by 2050) and a series of *Carbon-plus* scenarios (80% reduction in UK energy CO_2 emissions by 2050).

As a national model, UK M-M considers the detailed characteristics of the UK energy system. However, the role of international drivers is crucial for the quantification of the costs and pathways of a future LCS. The relative importance of international drivers (resources prices, technology learning, sustainable biomass, international aviation, trading of emission reductions) is explored through a transparent set of assumptions on each driver. The interactions between these international drivers are investigated through two combination scenarios: *Annex I consensus* and *Global consensus*.

An $100/$tCO_2$ carbon price delivers absolute reductions of around 55% in UK energy CO_2 emissions at a GDP loss of –0.36% in 2050. Moving to an 80% CO_2 reduction scenario (*Annex I consensus*) represents a considerable increase in GDP loss of –1.64% in 2050 and marginal CO_2 price of $402/$tCO_2$. This steep convexity in deep CO_2 reduction costs represents the growing technical and behavioural challenges associated with increasing efforts to decarbonize the UK energy system.

When considering the range of international drivers, for economy-wide results, the inclusion of international aviation and the potential large-scale purchases of CO_2 emissions reduction are most important. However, when all countries are meeting CO_2 reduction targets under LCS strategies (*Global consensus*), the availability, and hence cost reduction impact, of international purchases is drastically reduced.

When considering sectoral implications, all the international drivers considered here are important in both the trade-off between abatement in different upstream and downstream sectors (i.e. requiring both technological and behavioural change), and in the detailed configuration of specific sectors. An example of the latter point is the electricity sector, where small differences in levelized costs between the major zero-carbon options (coal and gas CCS, nuclear, intermittent renewables) lead to differing overall sizes and divergent configurations of the UK electricity system.

Combining these international drivers (*Global consensus*) leads to reduced costs in intermediate (2010–2030) periods (due to the impact of lower fossil prices and availability of emission purchases) but increased costs by 2050 due to the burden of international aviation. In 2050, costs rise to a GDP loss of 2.21% and a marginal CO_2 price of $589/$tCO_2$.

Although these scenarios of the costs and technological implications of the UK moving towards an LCS are illustrative, no systematic attempt is made to quantify the significant uncertainties in long-term pathways. However, UK policy makers should be cognisant of, and flexible with respect to, international strategies on LCS and emission reduction targets, and the impacts of key drivers on the UK energy sector. Post-Kyoto international mechanisms are one opportunity to influence and cooperate in LCS pathways.

Notes

1. See www.defra.gov.uk/environment/climatechange/internat/g8/index.htm.
2. Hydro, wastes, biomass, solar, wave, tidal, onshore wind, offshore wind, micro wind, nuclear and CCS (coal and natural gas) technologies.
3. United Nations Framework Convention on Climate Change.
4. Note that, because of increased aircraft efficiency, this still entails an increase of 105% in passenger km from 2000 to 2050. However, this would still be a significant departure from current trends, on the basis of which aviation demand would be projected to grow to two or three times its current level by 2030 (DTI, 2006).

5. The long run is defined as a 20-year period for capital stock to turn over. All costs are denoted in $2000.
6. For simplicity, only seven supply steps per MACC were assumed.
7. '... the use of the [Kyoto project] mechanisms shall be supplemental to domestic action and ... domestic action shall thus constitute a significant element of the effort made by each Party...' (DEFRA, 2007).
8. An alternative process would be to assume a perfectly competitive trading market, with some countries and regions undertaking substantially more emissions reductions (e.g. Russ and Criqui, 2007).
9. For example, the Renewables Obligation (15% renewable electricity generation by 2015), and the Energy Efficiency Commitment (EEC) in the industrial, services and residential sectors (Strachan et al., 2007).
10. Alternatively LCS-BSAV where the international aviation sector has been included.
11. An exchange rate of £1 = $1.8 is assumed.
12. However, in earlier years (2020 and 2030) when permits are available, domestic reductions are less.
13. From fossil fuel production, refining and distribution.
14. Resource costs entail a larger percentage of gas-fired generation costs than is the case for coal plants.
15. Due to continued restructuring of the UK economy, with improved energy intensity/efficiency of the energy sector (Strachan et al., 2007).

References

Contaldi, M., Gracceva, F., Tosato, G., 2007, 'Evaluation of green certificates policies using the MARKAL-Macro Italy model', *Energy Policy* 35, 797–808.

DEFRA, 2007, *Draft Climate Change Bill: Consultation Document*, Department for Environment, Food and Rural Affairs, London, March 2007.

DfT, 2004, *Aviation and Global Warming*, Department for Transport, London.

DTI, 2006, *Updated Energy Projections (EP-68)*, Department of Trade and Industry, London.

DTI, 2007, *Energy White Paper: Meeting the Energy Challenge*, Department of Trade and Industry, London [available at www.berr.gov.uk/energy/whitepaper/page39534.html].

EIA, 2007, *International Energy Outlook 2007*, US Energy Information Administration [available at www.eia.doe.gov/oiaf/ieo/index.html].

Ellerman, D., Decaux, A., 1998, *Analysis of Post-Kyoto CO_2 Emissions Trading using Marginal Abatement Curves*, Report 40, MIT Joint Program on the Science and Policy of Global Change.

European Commission, 2005, *World Energy Technology Outlook-2050 (WETO-H2)*, DG Research (Energy) [available at http://ec.europa.eu/research/energy/gp/gp_pu/article_1257_en.htm].

G8 Communiqué, 2007, *Chair's Summary*, G8 Heiligendamm Summit, 8 June 2007 [available at www.g-8.de/Webs/G8/EN/G8Summit/SummitDocuments/summit-documents.html].

Kannan, R., Strachan, N., Pye, S., Balta-Ozkan, N., 2007, *UK MARKAL Model Documentation* [available at www.ukerc.ac.uk/ResearchProgrammes/EnergySystemsandModelling/ESM.aspx].

Klepper, G., Peterson, S., 2006, 'Marginal abatement cost curves in general equilibrium: the influence of world energy prices', *Resource and Energy Economics* 28, 1–23.

Loulou, R., Goldstein, G., Noble, K., 2004, *Documentation for the MARKAL Family of Models* [available at http://www.etsap.org/index.asp].

Manne, A., Wene, C.-O., 1992, *MARKAL-Macro: A Linked Model for Energy–Economy Analysis*, Brookhaven National Laboratory, Report BNL-47161.

McDonald, A., Schrattenholzer, L., 2002, 'Learning curves and technology assessment', *International Journal of Technology Management* 23(7/8), 718–745.

Russ, P., Criqui, P., 2007, 'Post-Kyoto CO_2 emission reduction: the soft landing scenario analysed with POLES and other world models', *Energy Policy* 35(2), 786–796.

Sands, R., 2004, 'Dynamics of carbon abatement in the Second Generation Model', *Energy Economics* 26, 721–738.

Strachan, N., Kannan, R., Pye, S., 2007, *Final Report on DTI-DEFRA Scenarios and Sensitivities using the UK MARKAL and MARKAL-Macro Energy System Models*, Final Report for the Department of Trade and Industry, May 2007.

Weyant, J., 2004, 'Introduction and overview: EMF 19 study on technology and climate change policy', *Energy Economics* 26, 501–515.

WWF International, 2007, *A First Estimate of the Global Supply Potential for Bio-energy*, Briefing study commissioned by World Wildlife Foundation (WWF), July 2007.

climate
policy

■ research article

Effects of carbon tax on greenhouse gas mitigation in Thailand

RAM M. SHRESTHA*, SHREEKAR PRADHAN, MIGARA H. LIYANAGE

School of Environment, Resources and Development, Asian Institute of Technology, PO Box 4, Klong Luang, Pathumthani 12120, Thailand

This study analyses energy system development and the associated greenhouse gas emissions in Thailand under a reference case and three different carbon tax scenarios during 2013–2050 using a bottom-up cost-minimizing energy system model based on the Asia–Pacific Integrated Assessment Model (AIM/Enduse) framework. It considers the role of the renewable energy technologies as well as some emerging GHG-mitigating technologies, e.g. carbon capture and storage (CCS) in power generation, and GHG reduction in the country, and found that the power sector will play a major role in CO_2 emission reduction. Under the carbon tax scenarios, most of the CO_2 emission reduction (over 70%) will come from the power sector. The results also indicate the very significant potential for CO_2 emission reduction through a significant change in the transport system of the country by shifting from low-occupancy personal modes of transport to electrified MRTS and railways.

Keywords: carbon pricing; CO_2 reductions; energy systems; low-carbon society; policy instruments; scenario modelling; Thailand

Cette étude analyse le développement du système énergétique et les émissions de gaz à effet de serre associées en Thaïlande selon un cas de référence et trois différents scénarios de taxe carbone pendant la période 2013–2050 au moyen d'un modèle ascendant de minimisation du coût du système énergétique sur la base du cadre du modèle d'évaluation intégrée de l'Asie pacifique (AIM/Enduse). L'étude prend en compte le rôle des technologies d'énergie renouvelable ainsi que certaines technologies émergentes de réduction des GES, telles que la capture et le stockage du carbone (CSC) dans la production d'énergie et la réduction de GES dans le pays et révèle que le secteur de l'énergie jouera un rôle majeur dans la réduction des émissions du CO_2. Selon les scénarios de taxe carbone, la plupart des réductions d'émission de CO_2 (plus de 70%) viendront du secteur de l'énergie. Les résultats indiquent aussi un potentiel de réduction des émissions de CO_2 très caractéristique dans le système national du transport par le passage d'un mode de transport individuel à faible occupation au MRT et au ferroviaire.

Mots clés: fixation du prix du carbone; instruments de politique; modélisation de scenarios; réductions de CO_2; société sobre en carbone; systèmes énergétiques; Thaïlande

1. Introduction

Thailand is the second largest economy among the countries in the Association of South East Asian Nations (ASEAN) (IMF, 2008). The country is also the second-largest emitter of CO_2 in the ASEAN region. The CO_2 intensity of the country in 2004 was nearly 2.2 times that of the OECD as a whole. With the economy growing at over 5% per annum and increasing urbanization, the CO_2 emission in the country is expected to grow significantly in the future. In the face of global warming and growing international efforts to reduce greenhouse gas (GHG) emissions, it is

■ *Corresponding author. E-mail: ram@ait.ac.th

CLIMATE POLICY 8 (2008) S140–S155

doi:10.3763/cpol.2007.0497 © 2008 Earthscan ISSN: 1469-3062 (print), 1752-7457 (online) www.climatepolicy.com

important to identify the major options for GHG emission mitigation and their potential not only in industrialized countries but also in fast-growing developing countries such as Thailand.

Thailand is heavily dependent on imported energy in that it accounts for about 49% of the total primary energy supply in the country. Thus it is also of interest to analyse the effects of GHG reduction policy options on the development of a low-carbon economy and the energy security of the country over the longer term.

There have been several studies on energy system development and GHG emissions in Thailand (see, e.g., Shrestha et al., 1998, 2007; NEPO, 1999; Tanatvanit et al., 2003, 2004; Limmeechokchai and Suksuntornsiri, 2007a, 2007b). However, these studies do not examine the implications of carbon tax. Santisirisomboon et al. (2001) have analysed the effects of a carbon tax in the power generation sector of Thailand, while Malla and Shrestha (2005) analysed the effects of carbon tax on the Thai economy using a general equilibrium framework for the period 2000–2030. However, neither of these studies considered the effects of a carbon tax on disaggregated energy technology options for GHG reduction on the national energy system as a whole. Furthermore, these studies did not have a planning horizon covering the period up to 2050.

In the present study, we examine the prospects for CO_2 reduction from the Thai economy during 2013–2050 under three different carbon tax scenarios using a bottom-up energy system optimization model. The model includes different energy resource and technology options to meet the demand for energy services during the planning horizon of 2000–2050. In order to provide a broad range of options for GHG mitigation in the country, the model includes emerging technology options such as carbon capture and storage (CCS) along with clean coal technologies (such as IGCC, PFBC), fuel cells and hybrid vehicles, as well as conventional technology options in different sectors. It also includes biofuels as an energy resource option in the transport sector and the nuclear technology option for power generation.

Given the present heavy dependence of the country on low-occupancy vehicles as a means of passenger transport,[1] an analysis was also carried out to assess the effects of a partial shifting of the passenger transport demand from low-occupancy vehicles to the electrified mass rapid transport (MRT) system and railway services. This provides an insight into the level of GHG emission reduction that could be achieved through such policies in addition to the level of reduction achievable through a carbon tax in the absence of such modal shifts.

Section 2 describes the methodology used in the study; Section 3 provides scenario descriptions; the base case analysis is presented in Section 4, followed by analyses of the effects of carbon taxes in Section 5. The final section presents key conclusions and final remarks as well as the policy implications of the study.

2. Methodological approach

The study uses a bottom-up energy system model developed for Thailand using the Asia–Pacific Integrated Assessment Model (AIM)/Enduse framework (Kainuma et al., 2003). The model is broadly classified into two main components: (i) energy supply and conversion and (ii) service demand. The energy supply and conversion component represents energy extraction, imports and conversion of primary energy to secondary energy. In this component, coal mining, natural gas extraction, refining of crude oil, and power generation are considered. For power generation, 28 existing and new technology options are considered. Among the new technology options considered, seven are coal- and natural-gas-based carbon capture and storage (CCS) technologies. We have also considered nuclear as a potential power generation technology of the future.

The Thai economy is divided into five main sectors: agriculture, commercial, industrial, residential and transport. The industry sector has been subdivided into cement, steel, sugar, paper, chemicals, food, equipment, textiles and others. Similarly, transportation is sub-divided into passenger and freight transport. Passenger transportation is further divided into road, rail, air and water transport. All trading enterprises, hotels, restaurants, financial and telecommunication establishments are included in the commercial sector. The residential sector has been divided into urban and rural categories. Altogether 292 existing and candidate technology options are considered in the study for meeting end-use service demands. The future projections of service demands in agriculture, commercial, industrial and freight transport sectors are based on sub-sectoral value added, while the projection of service demands in the residential sector is based on number of households and appliance ownership per household. The service demand for passenger transport is projected based on population growth. Thailand's GDP projection during 2000–2016 is based on TDRI (2004), according to which the GDP would be growing at 6.4% by 2016. Thereafter, it is assumed that the GDP will grow at the rates of 6.4%, 5.3% and 4.5% per annum during 2016–2030, 2030–2040 and 2040–2050, respectively. On population, the medium variant forecast of the UN (2004) is considered in the model. It is assumed that service demand in a given year is linearly proportional to the value added in the year.

The model is based on linear programming and comprises an objective function to minimize total energy system cost year by year, subject to a number of constraints including those on service demand, energy resource availability, existing device stock, maximum allowable quantity of devices and emissions (see Kainuma et al., 2003 and NIES, 2007). The total cost comprises annualized fixed cost of recruited devices during a year, variable operating costs (i.e. operation and maintenance costs of devices, and fuel costs), cost of installing removal devices (e.g. flue gas desulfurizers for pulverized coal-fired power plants) and taxes.

3. Scenario descriptions

The base case scenario is defined as the business-as-usual case, i.e. the continuation of current economic, demographic and energy sector trends and policies, without any mitigation policy. The maximum availability of domestic fossil fuel resources (coal, oil and natural gas) during 2000–2050 under the base case is given in Table 1. However, no limit is imposed on imports of these fossil fuel resources.

TABLE 1 Energy resources reserve of Thailand (as of December 2005)

Type	Proven reserves	Probable reserves	Possible reserves	Total
Crude oil (million bbl)	192	119	76	387
Condensate (million bbl)	261	293	158	712
Natural gas (bcf)	10,743	11,598	9,555	31,896
Lignite (million tons)				2,870
Domestic hydro potential (MW)				15,112

Source: DEDE (2006a).

According to the latest Power Development Plan (PDP) (EGAT, 2007), 4,000 MW will be provided by nuclear, which is introduced from 2020. Thus the nuclear power generation option is also included in the study from 2020 onwards. The Thai Ministry of Energy has estimated a minimal amount of renewable energy resource potential for power generation at present in the country (Greacen and Bijoor, 2007). Thus the maximum exploitable level of solar and wind for power generation is assumed to be 5,000 MW and 1,100 MW, respectively, in this study. Likewise, the maximum exploitable level of agricultural residues (sugar cane residues – i.e. bagasse and tops and leaves, paddy husks, corncobs and others) is assumed to be 14,112 ktoe (Santisirisomboon et al., 2001; DEDE, 2006b; Prasertsan and Sajjakulnukit, 2006). The maximum level of plantation-based biomass is assumed to be 7,500 ktoe (Sajjakulnukit and Verapong, 2003; Santisirisomboon et al., 2001). It is assumed that the plantation-based biomass is produced on a sustainable basis and therefore there would be no net CO_2 emission involved. No carbon tax is considered in the base case scenario. The planning horizon of the study is 2000–2050 (50 years) with 2000 as the base year. A 10% discount rate (ADB, 1998) is used in the study.[2] All costs are expressed at the constant prices of the year 1995.

Three carbon tax scenarios are considered in the study.[3] They are:

- The use of a carbon tax of US$10/$tCO_2$ starting from 2013, which will exponentially rise to US$100/$tCO_2$ by 2050, all other things remaining the same as in the base case (hereafter called the 'C10+ scenario')
- The use of a carbon tax at a constant rate of US$75/$tCO_2$ from 2013 to 2050, all other things remaining the same as in the base case (hereafter, 'C75 scenario').
- The use of a carbon tax at a constant rate of US$100/$tCO_2$ from 2013 to 2050, all other things remaining the same as in the base case (hereafter, 'C100 scenario').

The carbon tax profiles in the three scenarios are shown in Figure 1.

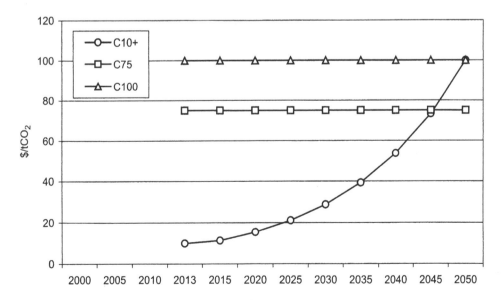

FIGURE 1 Three alternative carbon tax cases, $/$tCO_2$.

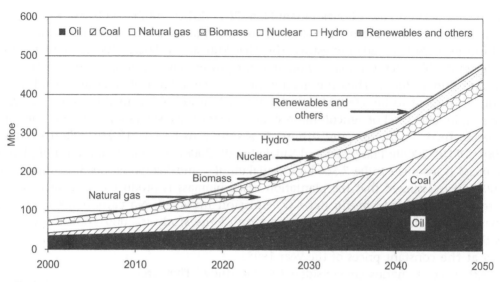

FIGURE 2 Total primary energy supply, Mtoe.

4. Base case analysis

4.1. Total primary energy supply (TPES) and energy mix

The total primary energy supply (TPES) increases with an annual average growth rate (AAGR) of 4.1% during 2000–2050. As a result, the TPES is found to increase from 75 Mtoe in 2000 to 484 Mtoe by 2050 (Figure 2). The combined share of oil and natural gas decreases from 71% in 2000 to 54% by 2050, while the share of coal (mainly imported) increases from 11% to 30% during the period. The use of nuclear power generation technology starts in 2020 and would acquire a share of 7% in TPES in 2050. The available biomass energy resource would be almost fully used from 2030 onwards; as a result, its share would decrease from 17% in 2000 to 7% in 2050. The share of hydroelectricity (including imported hydroelectricity) remains at about 1% throughout the period. Similarly, renewable energy (biogas, geothermal, solar and wind) and other sources (municipal solid waste) would have a very small share in TPES (less than 1%) during the study period. The results show that the share of imported energy (including imported coal, crude oil, natural gas, hydro and nuclear) will increase from 49% in 2000 to 74% by 2050.

4.2. Sectoral energy consumption

The total energy consumption (TEC) of different sectors is estimated to grow more than sixfold (i.e. from 47 Mtoe in 2000 to 376 Mtoe in 2050). The transport and industrial sectors dominate total energy consumption (see Table 2). Together these sectors accounted for 74% of TEC in 2000, and their share would grow to 84% by 2050. The share of industrial sector energy consumption grows till 2030 and decreases thereafter (Figure 3). The commercial sector energy consumption share would rise significantly, while the shares of the agriculture and residential sectors in TEC decline sharply during the planning horizon.

4.3. CO$_2$ emission

CO$_2$ emission is found to grow nearly sevenfold in the base case, i.e. from 158 million tonnes in 2000 to 1,262 million tonnes by 2050 (see Figure 4). Total cumulative CO$_2$ emission during the

TABLE 2 Share of sectoral energy consumption during 2000–2050 under base case, %

Sector	Year 2000	Year 2050
Transport	39	41
Industrial	35	43
Commercial	6	11
Residential	16	4
Agricultural	4	1
Total	100	100

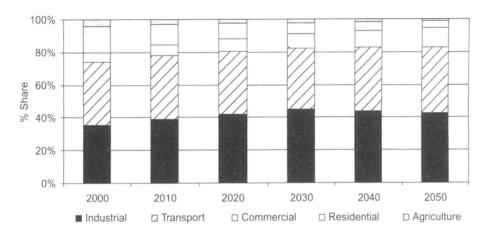

FIGURE 3 Yearly share of sectoral energy consumption during 2000–2050 under the base case, %.

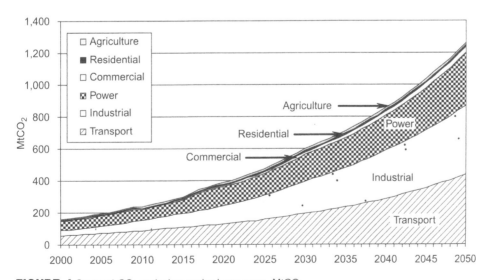

FIGURE 4 Sectoral CO_2 emission under base case, $MtCO_2$.

planning horizon is estimated to be 28,129 MtCO$_2$. The transport sector accounts for the highest share (34%) in cumulative CO$_2$ emission, followed by the industrial (33%) and power sectors (28%). Together, the transport, industry and power sectors had a share of over 93% of total CO$_2$ emission in 2000 and this would increase to 95% in 2050. The share of the power sector in CO$_2$ emission is found to decrease from 37% in 2000 to 26% in 2050. This is mainly because of penetration of nuclear energy technology and biomass-based power generation from 2020 onwards. The residential, agriculture and commercial sectors together have a relatively small share in CO$_2$ emission, which would decrease from 7% in 2000 to 5% by 2050.

5. Effects of carbon tax

The effects of three different levels of carbon tax (C10+, C75 and C100) on TPES, sectoral energy mix and CO$_2$ emission are discussed in the following sections.

5.1. Primary energy supply and fuel mix

The introduction of a carbon tax does not have a significant effect on the TPES. With the carbon tax, the TPES would increase from 75 Mtoe in 2000 to 471 Mtoe by 2050 under all the carbon tax scenarios. As a result of carbon tax, the share of natural gas is found to increase by 0%, 6% and 7% in scenarios C10+, C75 and C100, respectively, as compared with that in the base case, while the share of coal is found to decrease by 2%, 7% and 8% in the C10+, C75 and C100 scenarios, respectively, during 2000–2050. The share of natural gas in annual TPES is found to decrease until 2041 and would increase thereafter in C10+. In C75 and C100, the share of natural gas would decline until 2012, increase during 2013–2026, and would then decline again until 2050. The annual share of coal is found to increase gradually during the planning horizon (see Figure 5). The share of oil in cumulative TPES during the planning horizon would not change significantly, although the annual share of oil would decrease by 2022–2024 and would increase thereafter in all carbon tax scenarios. Thus, the carbon tax would have a major effect on the use of coal and natural gas, while its effect on oil would be negligible during 2000–2050.

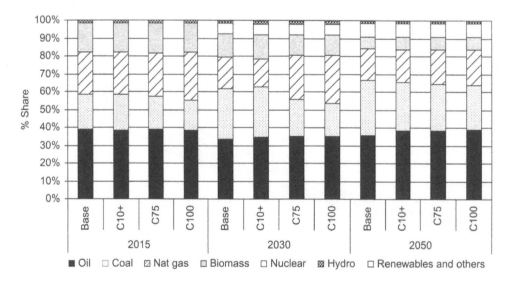

FIGURE 5 Fuel share in total primary energy supply in carbon tax scenarios, %.

5.2. Sectoral energy mix

The power, industrial and transport sectors are found to experience major changes in terms of the energy mix as a result of carbon tax. In the power sector, coal use would decrease while natural gas would increase significantly (Table 3). The nuclear generation share would not change during the planning horizon. The share of hydropower, biomass and other renewables would increase slightly with a carbon tax, while that of other renewables would have a small increment.

In the industrial sector, the share of coal in the cumulative energy consumption of the sector is found to decrease during the planning horizon with carbon tax (Table 4). The share of natural gas in the cumulative energy consumption of the sector during the planning horizon is found to increase over time. There would be no significant change in the shares of biomass, petroleum products and electricity.

In the transport sector, the shares of gasohol (90% gasoline) and biodiesel (90% diesel) vehicles in transport sector energy consumption would increase from nearly 5% in the base case to over

TABLE 3 Effect of carbon tax on energy mix of cumulative power generation during 2000–2050

Energy type	Percentage share in total energy used in the power sector			
	Base case	C10+	C75	C100
Coal	51	44	28	27
Natural gas	15	20	36	36
Biomass	7	8	8	8
Oil	0	0	0	0
Nuclear	15	15	15	15
Hydro	11	11	11	12
Other renewables	1	2	2	2

TABLE 4 Effect of carbon tax on energy mix (in cumulative energy consumption) of the industrial sector during 2000–2050

Energy type	Percentage share in total energy used in the industrial sector			
	Base	C10+	C75	C100
Biomass	8	7	7	7
Coal	41	40	36	34
Electricity	16	16	16	16
Natural gas	19	21	24	26
Petroleum products	9	9	9	9
Heat	7	7	7	7

TABLE 5 Effect of carbon tax on energy mix (in cumulative energy consumption) of the transport sector during 2000–2050

Type	Percentage share in total energy used in the transport sector			
	Base	C10+	C75	C100
Gasoline	49.7	21.6	22.7	22.8
Gasohol	1.4	13.8	13.9	13.9
Diesel	35.9	20.6	20.8	20.8
Biodiesel	3.4	38.7	39.0	39.0
Natural gas	7.9	3.7	2.1	2.0
LPG	1.4	1.3	1.2	1.2
Electric	0.1	0.1	0.1	0.1
Fuel cell	0.2	0.2	0.2	0.2

50% in all carbon tax scenarios during the planning horizon. On the other hand, the shares of pure gasoline and diesel vehicles in the sectoral energy consumption would fall under carbon tax scenarios. Because efficient gasohol and hybrid biodiesel vehicles are significantly more energy-efficient than natural gas vehicles, the shares of natural gas vehicles in the sector's energy consumption are found to fall under the carbon tax scenarios, while the shares of LPG, electric and fuel-cell vehicles are almost entirely unaffected (Table 5) .

5.3. CO_2 emission reduction

The CO_2 emissions over the planning period in the base case and the carbon tax cases are shown in Figure 6. Among the carbon tax scenarios, C100 results in the highest CO_2 emission reduction

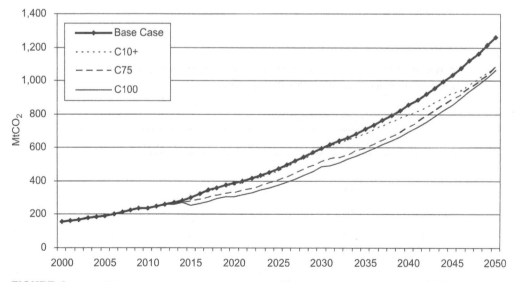

FIGURE 6 Yearly CO_2 emission under base case and different carbon tax scenarios, $MtCO_2$.

TABLE 6 CO_2 emission reduction in carbon tax scenarios

Sectors	Base case cumulative CO_2 emission, $MtCO_2$	Cumulative emission reduction from the base case emission level under carbon tax scenarios, $MtCO_2$		
		C10+	C75	C100
Agriculture	549	0	0	0
Commercial	712	0	0	0
Industrial	9,189	157	467	579
Residential	405	0	(8)	(6)
Transport	9,532	354	433	445
Power	7,743	1,179	2,810	3,607
Total	28,130	1,691	3,711	4,632

Note: The figure in parentheses denotes an increase in CO_2 emissions from the base case.

(16.5%) followed by C75 (13.2%) and C10+ (6.0%) (see Table 6). With carbon tax, the total system cost is found to increase by 6%, 16% and 20% in the C10+, C75 and C100 cases, respectively, during 2000–2050 as compared with that in the base case.

As a result of the carbon tax, the highest CO_2 emission reduction would take place in the power sector, followed by the transport and industrial sectors (see Table 6). The power sector accounts for 70% of total CO_2 emission reduction in the C10+ scenario, 76% in the C75 scenario and 78% in the C100 scenario. It should be noted here that, with the introduction of a carbon tax, reduction in CO_2 emission is mainly achieved through the use of coal-based carbon capture and storage (CCS) and natural-gas-based advanced combined cycle power generation.

In C10+, the share of the transport sector in CO_2 emission reduction is higher (21%) than that of the industrial sector (9%), while the opposite is the case in the C75 and C100 scenarios. The effect of carbon tax on CO_2 emission reduction in other sectors is not found to be significant.

In the agriculture sector, efficient electric motors and efficient diesel tractors were the only two types of efficient energy technology options considered in our model. Both these technologies would be selected in the base case to their maximum limit allowed. Thus, their usage would not be affected by carbon tax and, as a result, there is no reduction in CO_2 emission in this sector.

In the commercial sector, efficient air conditioners and compact fluorescent lamps (CFLs) were the efficient options considered for cooling and lighting service demands along with their conventional counterparts. It was found that CFLs would be selected to their maximum extent in the base case; thus carbon tax would not affect them. As for the efficient air conditioning devices, they would become cost-effective with carbon tax and would substitute for conventional air conditioning devices in the carbon tax cases. Although this reduces electricity consumption from the sector due to carbon tax, CO_2 emission reduction associated with the reduced electricity consumption is accounted for (included) under the power sector. Thus, there appears to be no reduction in CO_2 emission with the introduction of a carbon tax in the commercial sector.

In the residential sector, efficient lamps (fluorescent tubes and CFLs) would be selected to the maximum level in the base case; thus there is no change in their shares with the introduction of a carbon tax. In the case of space cooling service demands, as in the commercial sector, conventional air conditioners are substituted by efficient air conditioners under carbon tax cases; but the

CO_2 reduction associated with reduced electricity use by the efficient air conditioners is accounted for in the power sector CO_2 emission.

In residential cooking, electric stoves are partially replaced by efficient LPG stoves with the introduction of a carbon tax. With the increased use of LPG in the carbon tax cases, some increase in CO_2 emission would appear in the residential sector under the C75 and C100 scenarios (see Table 6). This is mainly because CO_2 emissions associated with electric stoves are reflected in the power sector emission rather than in the residential sector emission (see Table 6).

5.3.1. Role of renewable energy technologies in CO_2 emission reduction

In the base case, the share of renewable energy (including hydropower and biomass) in TPES is expected to decrease from 18% in year 2000 to 9% in year 2050. This is because biomass, hydropower and geothermal energy resources would be utilized to their maximum exploitable limit considered in the study after 2030 in the base case. In the power sector, the share of renewable energy in TPES would be 19% (including imported electricity and biomass) in the base case. The share would slightly increase to about 21% under carbon tax scenarios.

The use of solar photovoltaic (PV) technology in electricity generation has not been an attractive option in the base case, where the cost of solar PV is assumed to be fixed at US$4,240/kW (constant 1995 price) throughout the study period. However, the cost of solar technology is expected to fall over time due to the learning-by-doing effect (EPIA, 2006). IEA (2004) has used a learning rate of 18% for solar PV technology while analysing the competitiveness of CCS technology. Therefore, in this study we examined the effect of an 18% learning rate (LR) along with carbon tax. The results show that the learning-by-doing effect on solar technology would have significant effects on its adoption and CO_2 emission reduction in the power sector (Table 7).

5.3.2. Role of emerging energy technologies in CO_2 emission reduction

As discussed in Section 5.3, the carbon tax would have the largest effect on the power sector in terms of CO_2 emission reduction. One of the reasons for this is the adoption of a significant level of CCS-based power generation technologies under the carbon tax scenarios unlike in the base case, in which only a low level of CCS technology was cost-effective. The study shows that in the carbon tax scenarios, the share of electricity production based on CCS power generation technologies would increase to 7%, 14% and 23% in the C10+, C75 and C100 scenarios, respectively. Similarly the total power generation from natural gas power plants based on advanced combined cycle technology would increase from 14% in the base case to 19%, 35% and 31% in the C10+, C75 and C100 scenarios, respectively. Power generation from coal-fired IGCC and PFBC plants

TABLE 7 Effects of learning by doing on solar PV technology adoption and CO_2 emission reduction

	Base case	Carbon tax scenario (with 18% learning rate)		
		CT10+	CT75	CT100
Cumulative solar power generation, Mtoe	0	19	20	20
Power sector cumulative emission, $MtCO_2$	7,743	6,426	4,857	4,025
Emission reduction, $MtCO_2$	–	1,317	2,886	3,718
Year of penetration	–	2025	2013	2013

TABLE 8 Effects of increments in the CCS-based power plant costs on power generation and CO_2 emission from the power sector

	Base case	C10+		C75		C100	
		At reference price of CCS	At 25% higher price of CCS	At reference price of CCS	At 25% higher price of CCS	At reference price of CCS	At 25% higher price of CCS
Total power generation based on CCS technology, Mtoe	0	92	58	189	113	314	276
Total power generation based on combined cycle technology, Mtoe	199	266	275	483	532	434	472
Total CO_2 emission in the power sector, $MtCO_2$	7,742	6,563	6,737	4,932	5,247	4,135	4,248

would decrease from 26% in the base case to 19%, 1% and <1% in the C10+, C75 and C100 scenarios, respectively. Power generation from biomass-based IGCC plants would increase from about 2% in the base case to about 4% in the carbon tax scenarios.

The costs of the CCS type of power generation technologies used in this study are based on IEA (2004), while the costs for other power generation technologies are based on IEA (2001, 2005a, 2005b) and IAEA (2001). It should be noted that, being an emerging technology, the cost of CCS involves some uncertainties. Therefore we also examined the effect of a 25% increase in the capacity cost of CCS devices retrofitted to existing plants on CO_2 emission and level of CCS-based power generation under the carbon tax scenarios. As shown in Table 8, at the 25% higher cost of CCS capacity, the level of CCS-based power generation would be reduced, while power generation based on combined cycle power plants would increase. As a result, the power sector CO_2 emission would increase somewhat at the increased CCS capacity cost.

5.3.3. Effect of modal shift in passenger transport on CO_2 emission reduction

In Thailand, buses, cars, vans and pickups[4] are the most popular modes of passenger travel, along with motorcycles (OCLMT, 2000). In recent years the Bangkok Sky Train Service (BTS) and Mass Rapid Transit System (MRTS) have also been introduced, to a limited extent, for passenger transport services in the Bangkok metropolitan area (BTS, 2007; MRTA, 2007). It would therefore be of interest to investigate the effects on CO_2 emission in Thailand of shifting part of the passenger transport demand from low-occupancy passenger transport services (cars, vans and pickups) to MRTS and railway services from 2015 to 2050. For such an analysis, it is assumed that 10% of the passenger travel demand of cars, vans and pickups would be shifted to MRTS and railway services by the year 2015, while the shift would increase to 20% in 2030 and to 30% by the year 2050.[5] The results show that the modal shift in passenger transport would result in significant CO_2 emission reductions under the carbon tax scenarios (i.e. a reduction of 7.9% under C10+, 15.0% under C75, and 18.5% under C100) (see Table 9).

TABLE 9 Effects of modal shift in passenger transport on CO_2 emission reduction

Cumulative CO_2 emission in MtCO$_2$ during 2000–2050	Base	C10+	C75	C100
CO_2 emission without modal shift	28,130	26,439	24,419	23,498
CO_2 emission with modal shift		25,896	23,902	22,925
Total emission reduction		2,234	4,228	5,205

TABLE 10 Total NO_x and SO_2 emission reduction in carbon tax scenarios during 2000–2050

Sector	Base case NO_x emission Mtons	NO_x emission reduction, Mtons			Base case SO_2 emission Mt	SO_2 emission reduction, Mt		
		C10+	C75	C100		C10+	C75	C100
Industrial	27.8	0.7	1.9	2.5	65.8	9.0	18.9	17.7
Power	26.4	3.6	9.2	11.3	123.5	6.8	64.7	82.5
Transport	90.9	0.0	0.0	0.0	36.2	1.8	1.9	2.0
Others	10.0	0.0	0.1	0.0	5.6	0.0	0.1	0.0
Total	155.1	4.3	11.2	13.8	231.1	17.6	85.6	102.2

5.4. Co-benefits of carbon tax

Table 10 presents the effects of a carbon tax on SO_2 and NO_x emissions. The NO_x emission reduction would be the highest in the power sector, followed by the industrial sector (see Table 10). Similarly, the power sector accounts for the highest level of SO_2 emission reduction and is followed by the industrial and transport sectors. It should be noted that the reductions in SO_2 and NO_x emissions would be even larger if the aforementioned modal shift from low-occupancy vehicles to MRTs and railways is also considered.

6. Conclusions

This study shows that there would be almost a sevenfold increase in CO_2 emission in the base case during the planning horizon. Three sectors – i.e. transport, industry and power – together account for over 93% of the total CO_2 emission in 2000 and 95% by 2050.

With the introduction of a carbon tax, there would be a shift in energy mix from coal to natural gas in the power and industrial sectors. In the transport sector, the share of gasohol (90% gasoline) and biodiesel (90% diesel) use in vehicles would significantly increase, whereas the share of gasoline and diesel use in vehicles would decrease.

Under the C10+ scenario, the cumulative CO_2 emission reduction would be 6.0% during the planning horizon as compared with the base case emission. The corresponding figures under the C75 and C100 scenarios are 13.2% and 16.5%, respectively. If a modal shift of passenger transport from low-occupancy vehicles to MRTS and railways (from 10% in 2015 to 30% in 2050) was realized, the reduction in CO_2 emission would increase from 6.0% to 7.9 % in the C10+ case, from 13.2% to 15.0% in the C75 case, and from 16.5% to 18.5% in the C100 case.

The study also shows that the power sector will play a major role in CO_2 emission reduction. Most of the CO_2 emission reduction (over 70%) will come from the power sector under the carbon tax scenarios. This is mainly due to the adoption of CCS technologies as well as nuclear power generation technology. A sensitivity analysis with a 25% increase in the CCS capacity cost was found to marginally increase CO_2 emission from the power sector. Furthermore, the analysis reveals that biomass and other renewable energy technologies would not play a significant role in Thailand in CO_2 reduction under these carbon tax scenarios.

Thailand depends largely on imported energy sources and this dependence is expected to increase further in future. The imposition of a carbon tax does not seem to have any appreciable effect in reducing the energy import dependency of the country, due to its relatively small renewable energy resource potential. The results of the study also show that if Thailand is to pursue a development path towards a low-carbon society, CCS and nuclear technologies are the major options to be adopted for power generation in the country. Given the heavy reliance on low-occupancy personal vehicles for passenger transport, the results of the present study also show a very significant potential for CO_2 emission reduction through a significant change in the transport system of the country by shifting to electrified MRTS and railways from low-occupancy personal transport modes.

It should be noted that in the present study we have used a bottom-up partial equilibrium type of energy system model. As a result, the effects of a carbon tax on the demand for different types of energy have not been captured in our analysis. With the increase in the relative price of carbon-intensive fuels, the demand for services based on such fuels is expected to fall, which in turn would reduce the demand for such fuels. If such effects are considered in the model, a further reduction in CO_2 emission below the levels suggested by the present study is likely.

Acknowledgements

We would like to thank the two anonymous reviewers of this journal for their valuable comments and suggestions. However, we ourselves are solely responsible for any remaining errors in the article.

Notes

1. See OCLMT (2000).
2. The Asian Development Bank (ADB, 1998) states, citing the National Economic and Social Development Board (NESDB) of Thailand, that the Government of Thailand uses a 10% discount rate.
3. In the carbon tax and climate stabilization literature, studies have considered various carbon tax rates varying from US$10/tC to US$600/tC, with different intertemporal profiles (e.g. Edmonds et al., 2004; Smekens-Ramirez Morales, 2004; Vuuren et al., 2004). The three tax scenarios are well within the range of tax rates considered in the literature and also provide some variations in the tax profiles over time.
4. Vans and pickups are light-duty passenger vehicles having a carrying capacity of 12–16 people.
5. The passenger transport system in Thailand at present is predominantly road-based, with less than a 3% share for railways and MRTS. The government, in its policy document, has recently stated the goals of increasing the shares of railways and MRTS without giving any explicit targets. We have therefore considered these figures as a scenario for modal shift. Given that MRTS and railways already account for over 25% of the passenger transport services in some countries (e.g. Japan), the figures considered here as a scenario during 2015–2050 in the face of growing oil prices and climate change concern should not be considered unrealistic.

References

ADB, 1998, *Asia Least-cost Greenhouse Gas Abatement Strategy: Thailand*, Asian Development Bank, Global Environment Facility, United Nations Development Programme, Manila, Philippines.

BTS, 2007, *Bangkok Mass Transit Public Company*, Bangkok [available at www.mrta.co.th/eng/index.htm].

DEDE, 2006a, *Thailand Energy Situation 2005*, Department of Alternative Energy Development and Efficiency, Bangkok, Thailand.

DEDE, 2006b, *Biomass Potentials in Thailand*, Department of Alternative Energy Development and Efficiency, Ministry of Energy, Bangkok, Thailand [available at www.dede.go.th/dede/fileadmin/usr/berc/energysave/BiomassPotentialsinThailand.pdf].

Edmonds, J., Clarke, J., Dooley, J., Kim, S.H., Smith, S.J., 2004, 'Stabilization of CO_2 in a B2 world: insights on the roles of carbon capture and disposal, hydrogen, and transportation technologies', *Energy Economics* 26, 517–537.

EGAT, 2007, *Thailand Power Development Plan (PDP 2007)*, Electricity Generating Authority of Thailand, Bangkok, Thailand.

EPIA, 2006, *Solar Generation: Solar Electricity for over One Billion People and Two Million Jobs by 2020*, Greenpeace International and European Photovoltaic Industry Association (EPIA).

Greacen, C., Bijoor, S., 2007, 'Decentralized energy in Thailand: an emerging light', *World Rivers Review*, June 2007.

IAEA, 2001, *Greenhouse Gas Mitigation Analysis using ENPEP: A Modeling Guide*, International Atomic Energy Agency (IAEA), Vienna.

IEA, 2001, *Annual Energy Outlook 2001: Electricity Market Module*, International Energy Agency, USA [available at www.eia.doe.gov/oiaf/archive/aeo01/assumption/electricity.html].

IEA, 2004, *Prospects for CO_2 Capture and Storage: Energy Technology Analysis*, International Energy Agency (IEA) and Organisation for Economic Co-operation and Development (OECD), Paris.

IEA, 2005a, *Prospects for Hydrogen and Fuel Cells*, International Energy Agency (IEA) and Organisation for Economic Co-operation and Development (OECD), Paris.

IEA, 2005b, *Projected Costs of Generating Electricity*, International Energy Agency (IEA), Paris.

IMF, 2008, *World Economic Outlook Database, April 2008*, International Monetary Fund.

Kainuma, M., Matsuoka, Y., Morita, T. (eds), 2003, *Climate Policy Assessment: Asia–Pacific Integrated Modeling*, Springer, Tokyo.

Limmeechokchai, B., Suksuntornsiri, P., 2007a, 'Assessment of cleaner electricity generation technologies for net CO_2 mitigation in Thailand', *Renewable and Sustainable Energy Reviews* 11, 315–330.

Limmeechokchai, B., Suksuntornsiri, P., 2007b, 'Embedded energy and total greenhouse gas emissions in final consumptions within Thailand', *Renewable and Sustainable Energy Reviews* 11, 259–281.

Malla, S., Shrestha, R.M., 2005, 'Implications of carbon tax and energy efficiency improvement on Thai economy', paper presented at the Sixth IHDP Open Meeting 2005, Bonn.

MRTA, 2007, *Mass Rapid Transit Authority of Thailand* (MRTA), Bangkok, Thailand [available at www.mrta.co.th/eng/index.htm].

NEPO, 1999, *Thailand Energy Strategy and Policy: Final Report for Phase 1, Phase 2 and Phase 3*, National Energy Policy Office, Bangkok, Thailand.

NIES, 2007, *Aligning Climate Change and Sustainability: Scenarios, Modeling and Policy Analysis*, Center for Global Environmental Research, National Institute for Environmental Studies, Japan.

OCLMT, 2000, *Transport Data and Model Center: Final Report*, Office of the Commission for the Management of Land Traffic (OCLMT), Bangkok, Thailand.

Prasertsan, S., Sajjakulnukit, B., 2006, 'Biomass and biogas energy in Thailand: potential, opportunity and barriers', *Renewable Energy* 31, 599–610.

Sajjakulnukit, B., Verapong, P., 2003, 'Sustainable biomass production for energy in Thailand', *Biomass and Bioenergy* 25, 557–570.

Santisirisomboon, J., Limmeechokchai, B., Chungpaibulpatana, S., 2001, 'Impacts of biomass power generation and CO_2 taxation on electricity generation expansion planning and environmental emissions', *Energy Policy* 29(12), 975–985.

Shrestha, R.M., Khummonkol, P., Biswas, W.K., Timilsina, G.R., Sinbanchongjit, S., 1998, 'CO_2 mitigation potential of efficient demand-side technologies: the case of Thailand', *Energy Sources* 20, 310–316.

Shrestha, R.M., Malla, S., Liyanage, M.H., 2007, 'Scenario-based analyses of energy system development and its environmental implications in Thailand', *Energy Policy* 35, 3179–3193.

Smekens-Ramirez Morales, K.E.L., 2004, 'Response from a MARKAL technology model to the EMF scenario assumptions', *Energy Economics* 26, 655–674.

Tanatvanit, S., Limmeechokchai, B., Shrestha, R.M., 2004, 'CO_2 mitigation and power generation implications of clean supply-side and demand-side technologies in Thailand', *Energy Policy* 32, 83–90.

Tanatvanit, S., Limmeechokchai, B., Chungpaibulpatana, S., 2003, 'Sustainable energy development strategies: implications of energy demand management and renewable energy in Thailand', *Renewable and Sustainable Energy Reviews* 7, 367–395.

TDRI, 2004, GDP forecast 2000–2016. *Thailand Development and Research Institute* (TDRI), Bangkok, Thailand.

UN, 2004, *World Population Prospects: The 2004 Revision*, Population Division of the Department of Economic and Social Affairs of the United Nations Secretariat, New York [available at http://esa.un.org/unpp/].

Vuuren, D.P. van, de Vries, B., Eickhout, B., Kram, T., 2004, 'Responses to technology and taxes in a simulated world', *Energy Economics* 26, 579–601.

climate
policy

■ research article

Low-carbon society scenarios for India

P.R. SHUKLA[1]*, SUBASH DHAR[2], DIPTIRANJAN MAHAPATRA[3]

[1] Public Systems Group, Indian Institute of Management, Vastrapur, Ahmedabad 380015, India
[2] UNEP Risø Centre, Risø DTU, Frederiksborgvej 399, Building 142, Module 61, DK-4000 Roskilde, Denmark
[3] Indian Institute of Management, Vastrapur, Ahmedabad 380015, India

Low-carbon society scenarios visualize social, economic and technological transitions through which societies respond to climate change. This article assesses two paradigms for transiting to a low-carbon future in India. An integrated modelling framework is used for delineating and assessing the alternative development pathways having equal cumulative CO_2 emissions during the first half of the 21st century. The first pathway assumes a conventional development pattern together with a carbon price that aligns India's emissions to an optimal 550 ppmv CO_2e stabilization global response. The second emissions pathway assumes an underlying sustainable development pattern characterized by diverse response measures typical of the 'sustainability' paradigm. A comparative analysis of the alternative development strategies is presented on multiple indicators such as energy security, air quality, technology stocks and adaptive capacity, and conclusions are drawn.

Keywords: carbon pricing; climate change; development pathways; developing countries; energy mix; India; low-carbon society; scenario modelling; sustainable development

Les scénarios de sociétés sobres en carbone conçoivent des transitions sociales, économiques et technologiques à travers lesquelles la société répond au changement climatique. Ce papier évalue deux paradigmes de transition vers un futur sobre en carbone en Inde. Un cadre de modélisation intégrée est employé pour décliner et évaluer différents axes de développement ayant les mêmes émissions cumulatives de CO_2 pendant la première moitié du 21ème siècle. Le premier axe suppose un mode de développement traditionnel ainsi qu'un prix du carbone qui aligne les émissions de l'Inde à une réponse mondiale de stabilisation optimum de 550 ppmv CO_2eq. Le second axe adopte un mode de développement durable sous-jacent caricaturé par diverses mesures de réponse typiques du paradigme de « durabilité ». Une analyse comparative des différentes stratégies de développement est présentée sur la base d'indicateurs multiples tels que la sécurité énergétique, qualité de l'air, stocks de technologie et capacité d'adaptation et des conclusions sont tirées.

Mots clés: axes de développement; changement climatique; développement durable; fixation du prix du carbone; Inde; mix énergétique; modélisation de scénarios; pays en développement; société faiblement carbonée

1. Introduction

India faces major development challenges – access to the basic amenities such as drinking water, electricity, sanitation and clean cooking energy still remain a luxury for urban and rural dwellers alike (CoI, 2001).[1] Groundwater, which has been the key resource for meeting the irrigation and consumption needs of the urban and rural population, is coming under tremendous pressure because of haphazard urban planning and climate change (Burjia and Romani, 2003, cited in Mall et al., 2006). Environmental degradation in the future will have huge economic impacts on

■ *Corresponding author. E-mail*: shukla@iimahd.ernet.in

doi:10.3763/cpol.2007.0498 © 2008 Earthscan ISSN: 1469-3062 (print), 1752-7457 (online) www.climatepolicy.com

an agrarian and land-starved country such as India (Reddy, 2003).[2] Developing countries would need to build the adaptive capacity to face climate risks, with increasing evidence of climate change (IPCC, 2006). Climate change, due to an increase in greenhouse gas (GHG) emissions, is in turn related to increased human activities post-industrialization (IPCC, 2006), and therefore industrialization of large developing countries, such as China and India, can add significantly to GHG emissions. In the coming years, India faces challenges in economic development which have to be met with the limited resources available, with minimal externalities, and in the presence of large uncertainties with respect to climate.

One of the growing and accepted approaches to overcome this development paradox is through adoption of a sustainable development (SD) paradigm (Sathaye et al., 2006). SD is defined as 'development that meets the needs of the present without compromising the ability of future generations to meet their own needs' (WCED, 1987). The relationship between climate change and SD was recognized in the 'Delhi Declaration' during COP-8 in 2002 (Shukla et al., 2003). In fact, it has been argued that an exclusive, climate-centric vision would prove very expensive and might create a large mitigation and adaptation 'burden' (Shukla, 2006), whereas the SD pathway results in lower mitigation costs in addition to creating opportunities to realize the co-benefits without having to sacrifice the original objective of enhancing economic and social development (Shukla, 2006).

In this article we examine, using an integrated modelling framework, the realization of a low-carbon society through two alternative pathways. The first pathway uses a pure carbon policy instrument in the form of a carbon tax, whereas in the second we combine sustainable policies with a carbon tax.

2. Model framework

The integrated framework proposed in Figure 1 falls under the earlier AIM family of models (Kainuma et al., 2003; Shukla et al., 2004). In order to improve the policy interface, one new model – AIM-SNAPSHOT, which has a simple graphic interface, has been included. The bottom-up analysis is done by the MARKAL model (Fishbone and Abilock, 1981).

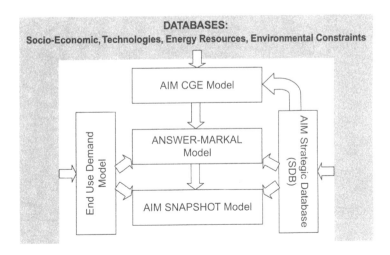

FIGURE 1 Integrated soft-linked model framework.

The need for a revised framework arose as the climate change discussion, backed up with increasing scientific evidence (IPCC, 2006), has become a more central and intensely debated topic among politicians and policy makers. The *Stern Review* and the *Energy Technology Strategies 2006* (IEA, 2006a; Stern, 2006) were a direct result of political mandates. In view of this, robust frameworks are required which convey to the policy makers, in simple terms, the impacts of alternative policies. The framework (Figure 1) uses the modelling resources developed over the last few years by the AIM team with a widely used energy system model ANSWER-MARKAL, and finally combines it with a model (SNAPSHOT model) that helps to present the results with adequate graphic interfaces.

2.1. Brief description of the component models

2.1.1. AIM-CGE

AIM-CGE is a top-down, computable general equilibrium (CGE), model developed jointly by National Institute of Environmental Studies (NIES), Japan, and Kyoto University, Japan (AIM Japan Team, 2005). The model is used to study the relationship between the economy and the environment (Masui, 2005). The top-down framework can perform cost analysis of both CO_2 mitigation and other GHG mitigation (Shukla et al., 2004). The model includes 18 regions and 13 sectors. The model can be used to assess the environmental and economic effects of new markets, new investment, technology transfer and international trade.

2.1.2. ANSWER-MARKAL model

MARKAL is a mathematical model for evaluating the energy system of one or several regions. MARKAL provides technology, fuel mix and investment decisions at detailed end-use level, while maintaining consistency with system constraints such as energy supply, demand, investment, emissions etc. A detailed discussion of the model concept and theory is provided at the ETSAP website (Loulou et al., 2004).

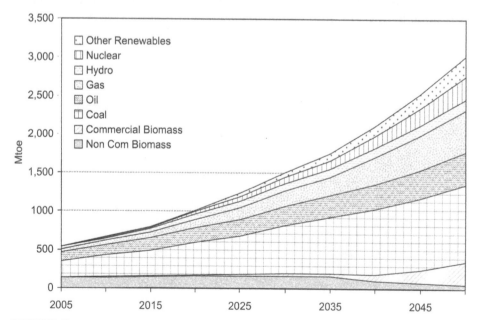

FIGURE 2 Fuel mix in the Base case scenario.

MARKAL has been used extensively for modelling the energy sector of India (Kanudia, 1996; Garg, 2000; Ghosh, 2000; Nair, 2003; TERI, 2006). ANSWER is the windows interface for the MARKAL model (ABARE, 1999).

2.1.3. End-use demand model

The Indian economy is presently on a high growth path: the demand for goods in the end-use sectors is showing high growth rates. The experience from developed countries has shown that these growth rates are likely to saturate as the economy modernizes. The approach used in the past is to model the demands using a logistic regression (Edmonds and Reilly, 1983; Grubler et al., 1993). First the long-term GDP projections are made using the historical data available (Ministry of Finance, 2007). Logistic regression using historical data is then used to estimate the sector-specific shares from industry, transport, commerce and agriculture. These sectoral shares when multiplied by GDP projections give us gross value added (GVA) for each sector. The last step involves estimation of the elasticity of each sub-sector (e.g. industry is divided into 11 sub-sectors such as steel, cement, etc.) with the sector-specific GVA. The elasticity is then used for estimating the future demand from each sector. The methodology described helps in capturing past trends and ensuring consistency with macroeconomic growth (Shukla et al., 2004).

2.1.4. AIM-SNAPSHOT model

The AIM-SNAPSHOT model is a spreadsheet tool designed to calculate the energy balance table and CO_2 emission table with inputs such as service demands, share of energy, and energy improvements by classifications of service and energy in the base and target years (NIES, 2006). The tool can be used for (i) developing and designing preliminary LCS and SD scenarios, (ii) performing 'what if' analysis, (iii) checking the consistency among the sectors, (iv) analysing the impacts of countermeasure packages, and (v) communicating with stakeholders.

2.2. Soft linking

The framework (Figure 1) contains a top-down model (AIM-CGE) which is soft-linked to a bottom-up model (ANSWER-MARKAL), which in turn is soft-linked to the AIM-SNAPSHOT model. Soft linking of models has been described in the literature (Bhattacharya et al., 2003; Nair et al., 2003). The inputs and outputs of each of the individual models are suitable for addressing specific but diverse economic, technological, social, environmental and energy sector issues, assuming consistent and similar assumptions and a shared database.

The top-down model, AIM-CGE, is used for estimating the GDP for different scenarios, and these are used as an exogenous input to the bottom-up ANSWER-MARKAL model. The ANSWER-MARKAL model provides detailed technology and sector-level energy and emission projections, which are in turn input to the AIM-SNAPSHOT model for factor analysis.

2.3. AIM strategic database (SDB)

Models require diverse databases such as economic growth, global and regional energy resource availability, input–output tables, sectoral and temporal end-use production processes and technologies, emission types, and much more. The data requirements are different for top-down and bottom-up models. The outputs from different models also serve as data for other models. There is essentially a complex flow of data between models and database whereby the models interact through the database in a soft-link framework. The AIM database plays a crucial role in ensuring data consistency across the models (Hibino et al., 2003; Shukla et al., 2004).

3. Scenario descriptions

The analysis considers three scenarios. The first scenario is the Base case, followed by two alternative pathways for achieving a low-carbon society (LCS). The scenario stories span the period up until 2050. The descriptions of the scenarios are below.

3.1. Base case scenario

This scenario assumes future economic development along the conventional path. In the case of a developing country, such as India, the scenario assumes the future socio-economic development to mirror the resource-intensive development path followed by the present developed countries. The assumptions about the key drivers – GDP, population and urbanization – are provided in Section 3.3. The annual GDP growth rate of 8% for the 27 years (2005–2032) is consistent with the moderate economic growth projections for India (GoI, 2006). The rate of population growth and urbanization follows the UN median demographic forecast (UNPD, 2006). The scenario assumes improvements in energy intensity similar to the dynamics-as-usual case (Shukla et al., 2003) and the targeted share of commercial renewable energy (Table 1.3, Shukla et al., 2003). This scenario assumes a stabilization target of 650 ppmv CO_2e. This would require CO_2 concentration stabilization at 550 ppmv, assuming that the contribution of non-CO_2 gases and land-use change is 100 ppmv CO_2e. The carbon price trajectory corresponds to stabilization at a 650 ppmv CO_2e concentration target or a 550 ppmv CO_2 concentration stabilization target for the CCSP SAP 2.1a equivalent scenario (Clarke et al., 2007).[3] The carbon price is $3/tCO_2$ during the Kyoto protocol period, rising to a modest $20/tCO_2$ in 2050 (Table 2).[4]

3.2. Low-carbon scenarios

3.2.1. Conventional path: Carbon tax (CT) scenario

This scenario presumes a stringent carbon tax (or permit price) trajectory compared with the milder carbon regime assumed under the Base case. As well as the difference in carbon tax, the underlying structure of this scenario is identical to the Base case. The scenario assumes a stabilization target of 550 ppmv CO_2e. The 550 ppmv CO_2e stabilization scenario translates to a CO_2 concentration stabilization at 480 ppmv, assuming that the contribution of non-CO_2 gases and land-use change is 70 ppmv CO_2e. The carbon price trajectory corresponding to stabilization at a 550 ppmv CO_2e concentration target is the same as that for the 480 ppmv CO_2 concentration stabilization target in the CCSP SAP 2.1a equivalent scenario. The carbon price trajectory for 480 ppmv CO_2 concentration stabilization, interpolated from CCSP SAP 2.1a stabilization scenarios (Clarke et al., 2007), is $10/tCO_2$ during the Kyoto Protocol period, rising to $100/tCO_2$ in 2050 (Table 2). The scenario assumes greater improvements in energy intensity and a higher target for the share of commercial renewable energy compared with the Base case scenario.

3.2.2. Sustainable society (SS) scenario
SUSTAINABILITY ALONE

This scenario represents a very different world view of development as compared with the Base case. The scenario follows a distinct 'sustainability' rationale, similar to that of the IPCC SRES B1 global scenario (IPCC, 2000). The scenario perspective is long term, aiming to deliver intergenerational justice by decoupling the economic growth from the highly resource-intensive and environmentally unsound conventional path. The storyline of the 'sustainability' scenario therefore cannot be constructed by starting with the Base case and making incremental changes.

The scenario rationale rests on aligning the economic development policies, measures and actions to gain multiple co-benefits, especially in developing countries where the institutions of governance, rule of law and markets are evolving. The scenario assumes the society to pro-actively introduce significant behavioural, technological, institutional, governance and economic measures which promote resource conservation (e.g. reduce, reuse, recycle), dematerialization, substitution among demands (e.g. information for transport), sustainable demographic transitions (e.g. in population growth and urbanization), urban planning, sustainable land use, efficient infrastructure choices (investments in alternative transport modes), innovations and technology transfer. The Indian 'sustainability' scenario also assumes a high degree of regional cooperation among the countries in southern Asia (Shukla, 2006) for energy and electricity trade and effective use of shared water and forest resources.

EMISSIONS STABILIZATION WITH SUSTAINABILITY

The 'emissions stabilization with sustainability' scenario assumes underlying socio-economic dynamics that are the same as in the 'sustainability alone' scenario. In addition, the scenario assumes a society which is responding to a globally agreed long-term CO_2 concentration stabilization target. The global target assumed for this analysis is a 550 ppmv CO_2e concentration target, or a temperature target in the range of 2° to 3°C. India's low-carbon society scenario then aims to generate carbon mitigation and adaptation responses to match the needs of a cost-effective 550 ppmv CO_2e global concentration stabilization regime. In comparison with the mild carbon tax assumed in the 'sustainability' scenario, the carbon price trajectory corresponding to the stabilization target is likely to be higher (Fisher et al., 2007), since it is explicitly responding to a carbon budget. Hence, India's cumulative CO_2 emissions (from 2005 to 2050) in the LCS scenario should be lower than in the 'sustainability' scenario. Instead of the Carbon tax trajectory, the SS scenario assumes a cumulative carbon budget for the post-Kyoto period 2013–2050, the rationale for which is discussed in Section 4.3.4. In comparison with the 'sustainability alone' scenario, the SS scenario will have higher penetration of decarbonization options such as carbon capture and storage (CCS) and new and renewable energy sources.

3.3. Scenario drivers

3.3.1. Macroeconomic

GDP for the period 2005–2032 is 8%; this is similar to the Planning Commission 8% GDP scenario (GoI, 2006). Population projections are based on the UN population medium scenario, version 2004, for India (UNPD, 2006). Population projections given by Census of India are only available until 2026 (CoI, 2006) and are therefore not used. The complete population and GDP assumptions are given in Table 1.

TABLE 1 Base case scenario drivers

Year	GDP (2005 prices)	Population	Period	Growth rate	
	(Billion Rs)	(Million)		GDP	Population
2005	32,833	1,103	2005–2030	8.1%	1.1%
2030	229,573	1,449	2030–2050	5.9%	0.5%
2050	774,673	1,593	2005–2050	7.1%	0.8%

3.3.2. Energy prices

A variety of prices are observed in the Indian energy markets, especially for coal and gas. The regulatory regime tries to keep prices aligned to the cost of production. Using the regulated prices information available in public domain, supply curves are created using a stepwise linear structure (Loulou et al., 2004).[5] The price assumptions for imported fuels are based on price projections given by IEA (IEA, 2006b).

3.3.3. Carbon prices

Carbon price trajectories for the Base case scenario and the Carbon tax scenario are linked to CO_2e stabilization targets of 650 and 550 ppmv CO_2e concentration, respectively. The price trajectories are obtained from outputs from the global second generation model (SGM) (J.A. Edmonds, personal communication, 2007). For the SS scenario, the price trajectory is similar to the Base case. However, India's cumulative carbon budget remains same as the cumulative emissions in the Carbon tax scenario. This cap on emissions results in a shadow price of carbon, which is given in Figure 12.

4. Results

The results presented in this article use an energy-accounting format which is different from that used by international agencies such as IEA whereby the contribution of renewables such as hydro, wind and solar resources to primary energy is equivalent only to the electricity generated. In contrast, for other resources such as fossil fuels and biomass, it is calculated in terms of the calorific value of the fuel. This depresses the share of hydro, wind and solar in primary energy and creates an inaccurate picture of the energy system. The current accounting practices (e.g. those followed by the IEA) also provide an undue incentive to use biomass over other renewables in the case of renewable energy targets (Larsen et al., 2007).

4.1. Base case: energy and emissions

The demand for energy increases 5.8 times to 3,016 Mtoe in 2050 as compared with 520 Mtoe in 2005, whereas the GDP increases by 23.6 times during the same period. Therefore, a decoupling of GDP and energy takes place as a result of changes in the structure of the economy and efficiency improvements. The energy intensity decreases at the rate of 3.29% for the period 2005–2050.

The energy mix diversifies from being highly dependent on coal, oil and traditional biomass to one which has a significant share of natural gas, other renewables, nuclear and commercial biomass. Diversification in terms of alternative forms of energy does not help in reducing the carbon intensities, as the share of renewables decreases from 29% in 2005 to 24% in 2050. However, the carbon intensities are moderated by an increase in the share of nuclear and gas at the expense of coal and oil, and therefore tend to follow the improvements in energy intensities (Figure 3).

The CO_2 emissions increase from 1,291 million tCO_2 in 2005 to 6,636 million tCO_2 in 2050. The cumulative emissions during this 45-year period are 162.3 billion tCO_2. The mild carbon tax trajectory (Table 2) is not adequate to bring in a pure carbon technology such as carbon capture and storage (CCS); however, it does lead to fuel substitutions. The fuel substitutions result in a faster reduction of carbon intensities as compared to energy intensities post-2040 (Figure 3).

TABLE 2 Carbon price trajectories (2005 US$/tCO$_2$)

Scenario	Base case (650 ppmv CO$_2$e)	Carbon tax (550 ppmv CO$_2$e)
Before 2012	3	10
2020	4	22
2030	8	40
2040	13	67
2050	20	100

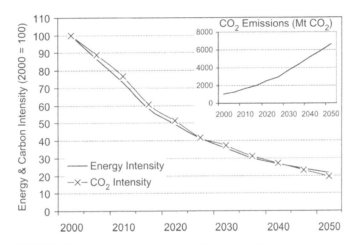

FIGURE 3 Energy and carbon intensities for the Base case scenario.

4.2. Carbon tax (CT) scenario

The Carbon tax scenario has a steep carbon tax trajectory (Table 2), which increases to US$100/tCO$_2$. To estimate the inefficiencies (Baumol and Blinder, 1999) and the resultant GDP loss, the AIM-CGE model was used. The tax revenues from the carbon tax are invested back into the economy. The GDP loss increases as the carbon tax increases and in 2050 the GDP is 1.35% lower than the Base case (Figure 4). These results are not significantly different from earlier studies that calculated GDP loss due to a carbon tax for India (Rana and Shukla, 2001).

The GDP loss, though not very significant, was used to recalculate the end-use demands for the Carbon tax scenario. The cumulative CO$_2$ mitigation for the period 2005–2050 comes to 62.6 billion tCO$_2$ and the mitigation happens mainly in the electricity sector (Figure 5), initially due to fuel switching. Post-2030, when the carbon prices exceed US$40/tCO$_2$, CCS along with coal-fired electricity generation, and CCS in steel and cement making also turn up as an option. The remaining mitigation happens due to a higher adoption of renewables, especially biomass, and improvements in device efficiencies.

The Base case scenario corresponds to 650 ppmv CO$_2$e stabilization, whereas the CT scenario corresponds to 550 ppmv CO$_2$e stabilization. The global emission trajectories show a very minor divergence between a 650 ppmv and a 550 ppmv CO$_2$e scenario up until 2050 (Edmonds et al., 2007). In India's case, there is a decoupling of CO$_2$ emissions post-2030 (Figure 6). The decoupling,

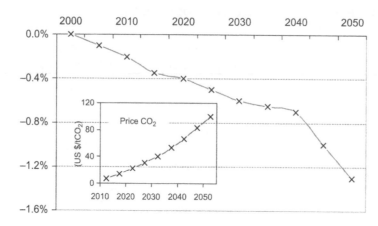

FIGURE 4 GDP loss in the Carbon price scenario.

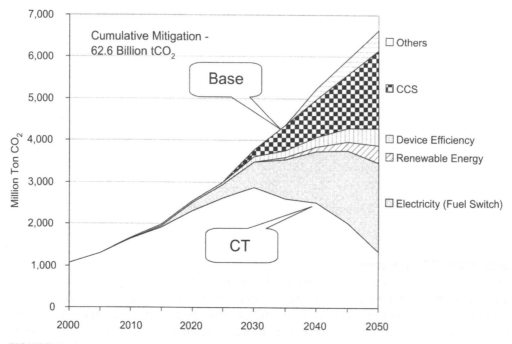

FIGURE 5 Mitigation options in the Carbon tax scenario.

however, takes 25 years, as there is an existing stock of energy infrastructures, and a large amount of investment in energy infrastructures for the future has already been committed. This indicates that the introduction of a tax and its impacts will have sufficient lags due to lock-ins.

4.3. Sustainable society (SS) scenario

4.3.1. Altering preferences and choices through policies

Policies for promoting sustainable development need to be based on the precautionary principle (Rao, 2000), as this helps in taking care of environmental unknowns (Myers, 1995). Therefore the emphasis is on reducing the anthropogenic influences, which are the root cause of GHG emissions,

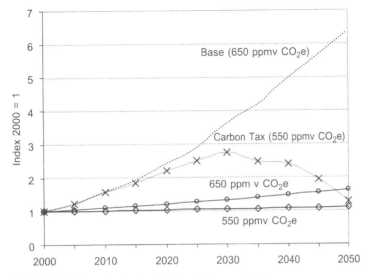

FIGURE 6 CO_2 emission trajectories India vs. global.

in all walks of life. However, the reduction of anthropogenic influences does not come at the expense of economic and social development and instead is committed to expanding the economic and climate frontiers (Shukla, 2005). The policies are frontier shifting through innovations in technology, institutions, international and regional cooperation, targeted technology and investment flows, aligning stakeholder interests, focusing on inputs (and not only outputs) and a long-term perspective to avoid lock-ins.

4.3.2. Demand projections

A sustainable society can arise through a number of policies, which eventually bring down the intermediate demand for products, while ensuring that the individual consumption of goods and services is not curtailed (the SS scenario assumes a GDP equivalent to the Base case). The final consumptions at an aggregate level may, however, come down owing to a reduction in population (the SS scenario assumes the UN low scenario for India; see UNPD, 2006).

The demand projections are done using sector-specific drivers which are changed keeping in mind the Sustainable society storylines. For example, for steel, the main consuming sectors are currently construction, capital goods and automobiles (Tata Steel, 2006). The impact of sustainability drivers on steel demand is given in Table 3.

TABLE 3 Impact of sustainable drivers on steel demand

Sector	Driver	Impact on steel demand
Transport	Urban planning Modal shift Substitution	Fewer automobiles, Less road transport infrastructure
Building	Building design Material substitution	More local materials, Low-rise buildings

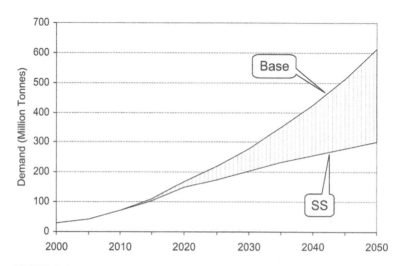

FIGURE 7 Carbon demand of steel industry 2000–2050 under the Base case and Sustainable society scenarios.

The steel demand would therefore be lower (Figure 7) as a result of sustainable policies. The final demand of energy from steel making may fall still lower because there is also greater recycling of steel in a sustainable society, and steel-making technologies are more efficient as compared with the Base case.

4.3.3. Back-casting approach

In the back-casting approach we first develop the emission target and then discuss the method for achieving this target (Kainuma et al., 2006). This approach is quite useful in the present case, as we want to keep the CO_2 mitigation the same between the Carbon tax and SS scenarios, in order to make the two scenarios comparable.

4.3.4. Emissions stabilization and SS

The approach to SS uses a sustainability paradigm; therefore in the first step we keep the tax trajectory similar to the Base case and introduce sustainability measures. The mitigation that happens by adopting the sustainability paradigm is of the order of 59.3 billion tCO_2. This is less than the 62.6 billion tCO_2 of mitigation that happens in the Carbon tax scenario. The CO_2 mitigation happens because of a diverse portfolio of measures across sectors. The transport sector accounts for a large share of mitigation, which happens due to modal shifts, reduced demand, and fuel switching. There is a reduction in demand from the industrial sector as demand diminishes for steel, cement and other energy-intensive commodities due to recycling, reuse, material substitutions, improvement of device efficiencies and fuel substitutions. The energy demand from agriculture is lower owing to reduced consumption due to improved agricultural practices related to irrigation and cropping patterns. Electricity demand, which is a derived demand, is also lower. Finally, there is an increasing reliance on renewable sources such as hydro, wind and solar.

The SS scenario, however, has mitigation equal to that in the CT scenario (Figure 8), and therefore a cap on cumulative emissions is kept, which is equivalent to the mitigation in the CT

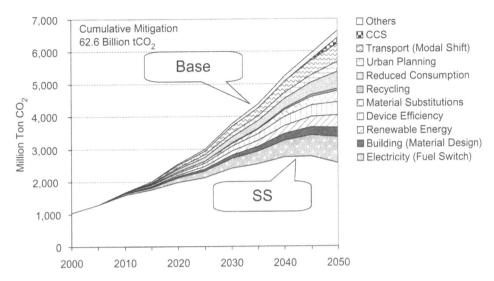

FIGURE 8 Mitigation options in the Sustainable society (SS) scenario.

scenario. The additional mitigation of 3.3 billion tCO_2 is achieved by additional fuel switching, improved device efficiency and using CCS. However, CCS accounts for just 0.5 billion tCO_2 mitigation as compared to the 19.1 billion tCO_2 that happens in the CT scenario.

4.4. Analysis of decarbonization

The decarbonization happens across scenarios as the energy intensities decline over time due to technological improvements and the changing structure of the economy, which result in a decoupling of economic growth and energy consumptions. The decline is faster in the case of SS (Figure 9) as the demand for intermediate goods and services reduces because of sustainable practices.

An alternative way of looking at two pathways for achieving a low-carbon society is in terms of consumption of final and intermediate goods and services. This is done by using the 'extended Kaya identity' (NIES, 2006). The change in CO_2 emissions from a base year is derived using the formula:

$$\text{Change in } CO_2 = \text{Demand effect } (D) + \text{Energy intensity effect } (E/D) + \text{Carbon intensity effect} \\ (C/E) + \text{Measures effect } (C'/C)$$

where D = driving forces (service demand of final and intermediate consumption), E = Energy consumption, C' = CO_2 emission without measures in the energy transformation sector, and C = CO_2 emission with measures in the transformation sector.

We observe (Figure 10) that there is a lower demand in the SS scenario. This lower demand comes primarily from intermediate sectors, e.g. the demand from the industrial sector increases by 462% in SS scenario as compared with 777% in the CT scenario. The energy intensities (Figure 9) with respect to demand are fairly similar in the two scenarios and, contrary to expectations, a sustainable society is more carbon-intensive. The overall level of emissions in the SS scenario (Figure 10) is also higher and therefore fewer countermeasures are required.

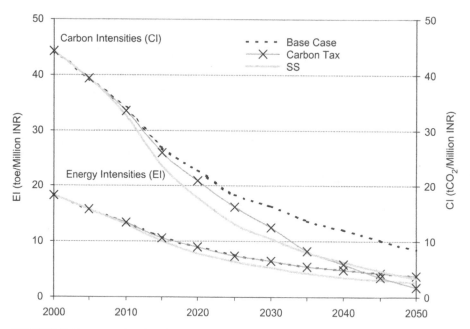

FIGURE 9 Energy and carbon intensities in the Base case and two low-carbon society (LCS) scenarios (Carbon tax and Sustainable society: SS).

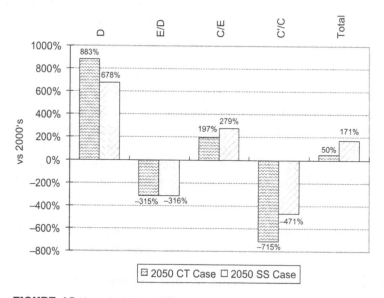

FIGURE 10 Kaya analysis of LCS scenarios.

4.5. Sustainability and carbon price

The CO_2 mitigation is the same between the sustainable and conventional or pure tax approaches. Note that the CO_2 emission pathways are different (Figure 11). The advance measures taken as a part of the sustainability paradigm place society on a low-carbon trajectory and, because of this, society can afford a higher level of emissions in 2050 as compared with the CT scenario (Figure 11).

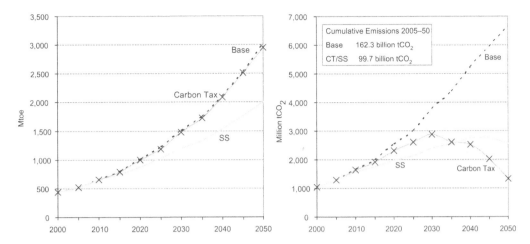

FIGURE 11 Energy and CO_2 emission trajectories across LCS scenarios.

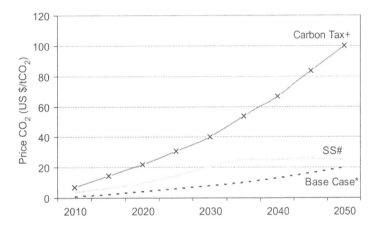

FIGURE 12 Carbon price in the LCS and Base case scenarios.

(*) Carbon price conforms to the global tax trajectory for 650 ppmv stabilization of CO_2e.
(#) Carbon price is the shadow price when for mitigation equivalent to CT scenario in the SS scenario.
(+) Carbon price conforms to the global tax trajectory for 550 ppmv stabilization of CO_2e.

In the SS scenario, major mitigation occurs due to sustainability measures (see Figure 8). This can have two implications. If the sustainability is restricted to a region (e.g. India), a higher mitigation corresponding to the global carbon price will occur, which can then be traded. If the sustainability paradigm is accepted globally, then a mild tax trajectory (Figure 12) is required.

5. Scenario comparisons: beyond carbon

5.1. Technologies, investment and institutional choice

The CO_2 mitigation choices differ between two LCS scenarios (Table 4). In the SS scenario, mitigation choices are more diverse and include measures that are designed to influence several development indicators simultaneously. The SS scenario pays greater attention to public investment decisions,

TABLE 4 Contributions to cumulative mitigations over Base case: 2005–2050 in billion tCO_2

Mitigation choice	SS	CT
Electricity (fuel switch)	13.4	30.5
Building (material design)	4.6	–
Renewable energy	6.2	2.8
Device efficiency	6.7	5.9
Material substitutions	4.9	–
Recycling	1.0	–
Reduced consumption	8.0	–
Urban planning	4.7	–
Transport (modal shift)	8.6	–
Others	3.8	4.3
CCS	0.5	19.1
Total mitigation	62.6	62.6

e.g. in infrastructure, which lead to modal shifts in the transport sector; and institutional interventions that alter the quality of development. In the CT scenario, the mitigation measures are more direct and have greater influence on private investments. In developing countries undergoing rapid transitions, aligning the development and carbon mitigation measures has significant advantages (Shukla, 2006).

In the CT scenario, where direct carbon mitigation technologies such as CCS find greater penetration, mitigation in sustainable society happens through diverse technology stocks. Implementing diversity of measures in SS would require building higher institutional capacity and influencing behaviours to reduce wasteful consumption as well as recycling and reuse of resources. In brief, in the SS scenario the mitigation is mainstreamed into the development pattern, causing a qualitative shift in the development vis-à-vis the Base case scenario. In the CT scenario, the mitigation actions take place at the margins of the economic development frontier.

5.2. Energy security

A major concern in the transition to a Low-carbon Society is its implications for 'Energy Security', i.e. the 'aggregate risk' related to energy vulnerabilities (Huntington and Brown, 2004), especially the energy supply (Correljé and van der Linde, 2006; Turton and Barreto, 2006) and its diversity (Dieter, 2002). In the Carbon tax scenario, the aggregate energy demand trajectory is almost identical to that in the Base case (Figure 11), whereas the energy demand is lower by a third in the SS scenario (Figure 13). The fossil fuel use declines in both LCS scenarios compared with the Base case scenario, although the CT scenario has significantly higher use of nuclear energy than the Base case scenario and a relatively higher use of fossil fuels together with a greater penetration of CCS technologies compared with the SS scenario. In the SS scenario, the dependence on oil, gas and nuclear energy reduces substantially. Since India has limited resource availability of these

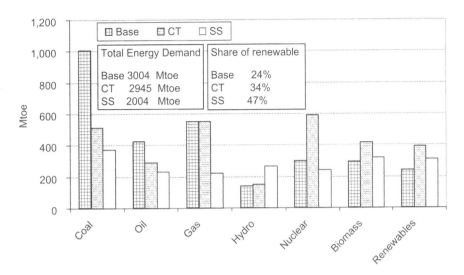

FIGURE 13 Fuel mix across the LCS and Base case scenarios (Year 2050).

fuels, the SS scenario will improve energy security in the conventional sense of dependence on energy imports (Yergin, 2006).

With regard to nuclear energy, the Base case scenario has a nuclear capacity of 178.5 GW in 2050. A fraction of this capacity corresponds to the conventional fuel cycle with dependence on imported uranium fuel. The remainder is the capacity under the three-stage nuclear programme, which would use indigenously available thorium as fuel (Kakodkar, 2006). The CT scenario has nuclear capacity of 326.4 GW in 2050. Compared to the Base case, the CT scenario requires higher import of uranium and this will adversely impact energy security.

5.3. Co-benefits of conjoint mitigation

Energy emissions contribute significantly to the local air quality in urban and industrial areas. The control of local air pollutants, e.g. SO_2, has been a major aim of environmental programmes in the developed world (Alcamo et al., 1990; Stavins, 1998; Ellerman et al., 2000). But at the time when SO_2 controls were initiated in the developed world, climate change was not yet a major concern. In India, where SO_2 control policies are being instituted more recently (Garg et al., 2006), there are opportunities to develop conjoint measures to control SO_2 and CO_2. Whereas the Base case scenario includes dynamics-as-usual SO_2 control measures which by themselves would decouple economic growth and the SO_2 emissions, the LCS scenarios would lead to higher and cheaper reduction in SO_2 emissions (Figure 14) since the conjoint measures would share the cost of their simultaneous mitigation. Thus, during the low-carbon transition, the conjoint policies can deliver benefits of improved air quality or alternatively through the reduced cost of achieving air quality targets. Evidently, the Sustainable society scenario would deliver greater air quality co-benefits than the Carbon tax scenario.

5.4. Adaptive capacity

Sustainable development is characterized by higher investment in human and social capital (Arrow et al., 2004) compared with that under conventional development. In developing countries this

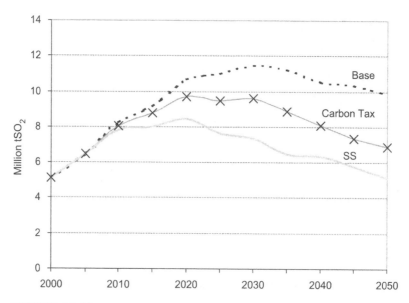

FIGURE 14 SO$_2$ emissions in the LCS and Base case scenarios.

translates into higher capabilities, especially among lower-income groups, to adapt to risks. Thus, a low-carbon society following a 'sustainability paradigm' would also deliver additional co-benefits vis-à-vis climate change risks.

There are several areas where mitigation and adaptation are directly linked. A prominent example is the dams which produce multiple outputs such as water supply for irrigation, electricity, flood control, and canals for shipping. Under conventional development, the public-good character of outputs such as 'flood control' and the inability to charge the full costs of infrastructure from the basic products such as water and electricity, especially in developing countries, leads to less than socially optimal deployment of hydro resources. In sustainable societies, the institutions are focused to overcome such failures. In addition, in southern Asia, there are Himalayan rivers shared by several countries. Regional cooperation institutions, the hallmark of sustainable societies, are expected to better recover the direct, indirect and spillover benefits of regional hydro resources and this will therefore lead to their higher deployment (Figure 13). Besides lowering carbon emissions, hydro projects would add to the adaptive capacity vis-à-vis enhanced climate variability (Kumar et al., 2003) of agriculture and population at risk from floods and droughts.

6. Conclusions: achieving LCS with sustainability

Many of India's development choices today and in the near future will determine its carbon emissions pathways for the long term. This article analysed two pathways for India's transition to a 'low-carbon society'. The pathways correspond to two different paradigms. The first, which follows the conventional development paradigm, treats carbon mitigation as an issue to be treated at the margin of development decisions through carbon-centric, market-efficient instruments such as a carbon tax or permits to decouple carbon emissions from the economy. This pathway has little direct implication for major development choices, including aggregate energy demand. The alternative paradigm considers low-carbon transition as an issue embedded within the larger development issue of transition to a 'sustainable society'. The strategy in this case is to mainstream carbon emissions mitigation by embedding low-carbon choices within the numerous development decisions. Thus, the low-carbon society transition through the 'sustainability' route decouples economic growth not only from carbon but also from several key resources, including energy. In this scenario, weaker carbon price signals would be an adequate driver for low-carbon transition. The mitigation signals would manifest through a diverse portfolio of technologies, with relatively little dependence on pure carbon mitigation technologies, such as CCS, which could have negative development dividends.

Renewable energy sources emerge as a preferred choice for carbon mitigation in both the Carbon tax scenario and the Sustainable society scenario, although their drivers are different. In the CT scenario, the relative price difference between renewable and fossil fuels is reduced by a carbon tax which enables faster penetration of renewables. In a sustainable society, the co-benefits of renewable energy, as well as higher deployable potential and lower transaction costs due to cooperation among the stakeholders, propel the penetration of renewable resources. Such a low-carbon transition would be accompanied by improved local environment and energy security, which are the key issues for a rapidly developing large economy like India. These issues would need to be addressed regardless of carbon mitigation.

Even in a low-carbon world, there will be significant climate change to which society must adapt. In a large developing country such as India, which would see a transition to higher incomes in the 21st century, the conventional path would exert enormous pressure on natural resources and ecosystems that could be large enough to impede global economic growth. In contrast, the prudent use of natural resources in the global 'sustainability' vision would reduce resource competition and conflicts, reduce prices of resources, and permit sustained higher economic growth. This, together with a greater emphasis on social and human capital under the sustainability vision, would increase the adaptive capacity to counter the adverse impacts of climate change.

Finally, in a globalizing world, a single country cannot decide a development pathway that is significantly different from the global trend. The advancement of knowledge stocks, on which the future technology transitions occur, depends on global efforts. In addition, the global cost-effectiveness of carbon mitigation requires the equalization of carbon prices across nations. In our analysis of a SS scenario for India, a significantly lower carbon mitigation shadow price is needed in order to achieve the same cumulative mitigation, compared with the carbon tax needed if the global as well as India's economic development followed a conventional path. A globally efficient low-carbon transition would require harmonization of development visions across nations. The sustainable global development, led by the industrialized nations, would thus be a precondition for sustainable development in developing countries, and also for aligning the low-carbon society transition in developing countries with their sustainable development goals.

Acknowledgements

We are grateful to Dr Jae Edmond for very insightful discussions on a 'global technology strategy' for transition to a low-carbon future and for providing carbon price data from the global CO_2e stabilization modelling runs. We are thankful to the National Institute of Environment Studies (NIES), Japan, for access to the Asia–Pacific integrated model (AIM) and the strategic database. We received very valuable inputs from Professor Yuzuru Matsuoka of Kyoto University, Dr Shuzo Nishioka, Dr Mikkio Kainuma, Dr Toshihiko Masui and Dr Junichi Fujino from NIES, and Mr Go Hibino from the Mizuho Information and Research Institute, Japan. We received important insights into low-carbon society scenarios from the two workshops organized under the 'Japan–UK Low-Carbon Society' project. Thanks are especially due to Mr Naoya Tsukamoto from the Ministry of

Environment, Japan and Dr David Warrilow from DEFRA, UK, for their encouragement and support. Above all, we wish to acknowledge numerous Indian researchers, policy makers, industry experts and NGOs for their cooperation in sharing valuable information and insights into the complex future transition processes underlying the scenario specifications and nuanced modelling.

Notes

1. Of the total 192 million households (as per the 2001 census) only 18% have access to modern cooking energy such as LPG and only 56% of households have access to electricity for lighting (CoI, 2001).
2. Reddy has estimated a GDP loss of 1.4% because of land degradation.
3. The US Climate Change Science Program Synthesis and Assessment Product 2.1a (CCSP SAP 2.1a) used three models: the integrated global systems model (IGSM), model for evaluating the regional and global effects (MERGE), and MiniCAM. Four GHG stabilization scenarios corresponding to CO_2 concentration levels of 450 ppm, 550 ppm, 650 ppm and 750 ppm were evaluated using the models (Clarke et al., 2007).
4. $ corresponds to the year 2005 US dollar.
5. Information related to coal prices can be obtained from the website of the Ministry of Coal, whereas information on oil and gas prices was taken from the Infraline database.

References

ABARE, 1999, *User Manual: ANSWER-MARKAL, an Energy Policy Optimisation Tool, Version 3.3.4*, Australian Bureau of Agricultural and Resource Economics, Canberra.

AIM Japan Team, 2005, *AIM-CGE [Country]: Data and Program Manual*, National Institute for Environmental Studies, Tsukuba, Japan.

Alcamo, J., Shaw, R., Hordijk, L., 1990, *The RAINS Model of Acidification: Science and Strategies in Europe*, Kluwer Academic Publishers, Dordrecht, The Netherlands.

Arrow, K., Dasgupta, P., Goulder, L., Daily, G., Ehrlich, P., Heal, G., Levin, S., Mäler, K.-G., Schneider, S., Starrett, D., Walker, B., 2004, 'Are We Consuming Too Much?', *Journal of Economic Perspectives* 18(3), 147–172.

Baumol, W.J., Blinder, A.S., 1999, *Economics: Principles and Policy*, 8th edn, Dryden Press, Fort Worth, TX.

Bhattacharya, S., Ravindranath, N.H., Shukla, P.R., Kalra, N., Gosain, A.K., Kumar, K.K., 2003, 'Tools for vulnerability assessment and adaptation', in: P.R. Shukla, S.K. Sharma, N.H. Ravindranath, A. Garg, S. Bhattacharya (eds), *Climate Change and India: Vulnerability Assessment and Adaptation*, Universities Press, Hyderabad, India.

Clarke, L.E., Edmonds, J.A., Jacoby, H.D., Pitcher, H.M., Reilly, J.M., Richels, R.G., 2007, 'Scenario of greenhouse gas emissions and atmospheric concentrations', in: *Synthesis and Assessment Product 2.1a*, United States Climate Change Science Program and Subcommittee on Global Change Research.

CoI (Census of India), 2001, *Census Data 2001: H – Series: Tables on Houses, Household Amenities and Assets*, Vol. 2007 [available at www.censusindia.gov.in/Census_Data_2001/Census_data_finder/Census_Data_Finder.aspx].

CoI (Census of India), 2006, *Population Projections for India and States 20012-026*, Report of the Technical Group on Population Projections constituted by the National Commission on Population, Office of the Registrar General and Census Commissioner, New Delhi, India.

Correljé, A., van der Linde, C., 2006, 'Energy supply security and geopolitics: a European perspective', *Energy Policy* 34, 532–543.

Dieter, H., 2002, 'Energy policy: security of supply, sustainability and competition', *Energy Policy* 30, 173–184.

Edmonds, J., Reilly, J., 1983, 'A long-term energy-economic model of carbon dioxide release from fossil fuel use', *Energy Economics* 5(2), 74–88.

Edmonds, J.A., Wise, M.A., Dooley, J.J., Kim, S.H., Smith, S.J., Runci, P.J., Clarke, L.E., Malone, E.L., Stokes, G.M., 2007, 'Global energy technology strategy: addressing climate change', in: *Phase 2 Findings from an International Public–Private Sponsored Research Program*, Joint Global Change Research Institute, Pacific Northwest National Laboratory and Battelle Memorial Institute, MD.

Ellerman, A.D., Joskow, P.L., Schmalensee, R., Montero, J.-P., Bailey, E.M., 2000, *Markets for Clean Air: The US Acid Rain Program*, Cambridge University Press, Cambridge, UK.

Fishbone, L.G., Abilock, H., 1981, 'MARKAL: a linear programming model for energy system analysis – technical description of the BNL version', *International Journal of Energy Research* 5, 353–375.

Fisher, B.S., Nakicenovic, N., Alfsen, K., Corfee Morlot, J., de la Chesnaye, F., Hourcade, J.-C., Jiang, K., Kainuma, M., La Rovere, E., Matysek, A., Rana, A., Riahi, K., Richels, R., Rose, S., van Vuuren, D., Warren, R., 2007, 'Issues related to mitigation in the long-term context', in: B. Metz, O.R. Davidson, P.R. Bosch, R. Dave, L.A. Meyer (eds), *Climate Change 2007: Mitigation*. Contribution of Working Group III to the Fourth Assessment Report of the Intergovernmental Panel on Climate Change, Cambridge University Press, Cambridge, UK.

Garg, A., 2000, 'Technologies, policies and measures for energy and environment future', Doctoral dissertation, Indian Institute of Management, Ahmedabad, India.

Garg, A., Shukla, P.R., Kapshe, M., 2006, 'The sectoral trends of multigas emissions inventory of India', *Atmospheric Environment* 40(24), 4608–4620.

Ghosh, D., 2000, 'Long-term technology strategies and policies for Indian power sector', Doctoral dissertation, Indian Institute of Management, Ahmedabad, India.

GoI (Government of India), 2006, *Integrated Energy Policy: Report of the Expert Committee*, Planning Commission, New Delhi, India.

Grubler, A., Nakicenovic, N., Schafer, A., 1993, *Dynamics of Transport and Energy Systems: History of Development and a Scenario for Future*, International Institute for Applied Systems, Laxenburg, Austria.

Hibino, G., Matsuoka, Y., Kainuma, M., 2003, 'AIM/common database: a tool for AIM family linkage', in: M. Kainuma, Y. Matsuoka, T. Morita (eds), *Climate Policy Assesment: Asia–Pacific Integrated Modeling*, Springer, Tokyo.

Huntington, H.G., Brown, S.P.A., 2004, 'Energy security and global climate change mitigation', *Energy Policy* 32, 715–718.

IEA, 2006a, *Energy Technology Perspectives 2006: Scenarios and Strategies to 2050*, OECD/IEA, Paris.

IEA, 2006b, *World Energy Outlook 2006*, OECD/IEA, Paris.

IPCC, 2000, *Emission Scenarios*, Cambridge University Press, Cambridge, UK.

IPCC, 2006, 'Summary for Policymakers', in: *Climate Change 2007: The Physical Basis*, Intergovernmental Panel on Climate Change, Geneva.

Kainuma, M., Matsuoka, Y., Morita, T., 2003, 'AIM modeling: overview and major findings', in: M. Kainuma, Y. Matsuoka, T. Morita (eds), *Climate Policy Assesment: Asia–Pacific Integrated Modeling*, Springer, Tokyo.

Kainuma, M., Masui, T., Fujino, J., Ashina, S., Matsuoka, Y., Kwase, R., Akashi, O., Hibino, G., Miyashita, M., Ehara, T., Pandey, R., Kapshe, M., Piris-Cabezas, P., 2006, *Development of Japan Low Carbon Scenarios*, Ministry of the Environment, Japan.

Kakodkar, A., 2006, 'Role of nuclear in India's power-mix', in: IRADe (ed), *Energy Conclave 2006: Expanding Options for Power Sector – Infraline Database* [available at www.infraline.com/power/default.asp?idCategory=2275&URL1=/power/Presentations/Others/EnergyConclave06/EnergyConclaveConferencePresent2006-Index.asp].

Kanudia, A., 1996, 'Energy–environment policy and technology selection: modelling and analysis for India', Doctoral dissertation, Indian Institute of Management, Ahmedabad, India.

Kumar, K.R., Kumar, K.K., Prasanna, V., Kamala, K., Deshpande, N.R., Patwardhan, S.K., Pant, G.B., 2003, 'Future climate scenarios', in: P.R. Shukla, S.K. Sharma, N.H. Ravindranath, A. Garg, S. Bhattacharya (eds), *Climate Change and India: Vulnerability Assessment and Adaptation*, Universities Press, Hyderabad, India.

Larsen, H., Sønderberg Petersen, L., 2007, 'Future options for energy technologies', in: *Risø Energy Report 6*, Risø Energy Centre, Roskilde, Denmark.

Loulou, R., Goldstein, G., Noble, K., 2004, *Documentation for the MARKAL Family of Models, October 2004*, Vol. 2007 [available at www.etsap.org/documentation.asp].

Mall, R.K., Gupta, A., Singh, R., Singh, R.S., Rathore, L.S., 2006, 'Water resources and climate change: an Indian perspective', *Current Science* 90(12), 1610–1626.

Masui, T., 2005, 'Concept of CGE model and simple GE model based on IO data', in: *AIM Training Workshop 2005*, National Institute of Environmental Studies, Tsukuba, Japan.

Ministry of Finance, 2007, *Economic Survey 20062-007*, Government of India, New Delhi, India.

Myers, N., 1995, 'Environmental unknowns', *Science* 269, 258–260.

Nair, R., 2003, 'Energy Security in South Asia: integrating the primary energy and electricity markets', Doctoral dissertation, Indian Institute of Management, Ahmedabad, India.

Nair, R., Shukla, P.R., Kapshe, M., Garg, A., Rana, A., 2003, 'Analysis of long-term energy and carbon emission scenarios for India', *Mitigation and Adaptation Strategies for Global Change* 8, 53–69.

NIES, 2006, *Energy Snapshot Tool (ESS): Manual*, National Institute of Environmental Studies, Tsukuba, Japan.

Rana, A., Morita, T., 2000, 'Scenarios for greenhouse gas emission mitigation: a review of modeling of strategies and policies in integrated assessment models', *Environmental Economics and Policy Studies* 3(2), 267–289.

Rana, A., Shukla, P.R., 2001, 'Macroeconomic models for long-term energy and emissions in India', *OPSEARCH* 38(1).

Rao, P.K., 2000, *Sustainable Development: Economics and Policy*, Blackwell Publishing, Boston, MA.

Reddy, R.V., 2003, 'Land degradation in India: extents, costs and determinants', *Economic and Political Weekly* 38(44), 4700–4713.

Sathaye, J., Shukla, P.R., Ravindranath, N.H., 2006, 'Climate change, sustainable development and India: global and national concerns', *Current Science* 90(3), 314–325.

Shukla, P.R., 2005, 'Aligning justice and efficiency in the global climate regime: a developing country perspective', in: W. Sinnott-Armstrong, R.B. Howarth (eds), *Perspectives on Climate Change: Science, Economics, Politics, Ethics*, Vol. 5, Advances in the Economics of Environmental Resources, Elsevier, Oxford, UK.

Shukla, P.R., 2006, 'India's GHG emission scenarios: aligning development and stabilization paths', *Current Science* 90(3), 384–395.

Shukla, P.R., Sharma, S.K., Garg, A., Bhattacharya, S., Ravindranath, N.H., 2003, 'Climate change vulnerability assessment and adaptation: the context', in: P.R. Shukla, S.K. Sharma, N.H. Ravindranath, A. Garg, S. Bhattacharya (eds), *Climate Change and India: Vulnerability Assessment and Adaptation*, Universities Press, Hyderabad, India.

Shukla, P.R., Rana, A., Garg, A., Kapshe, M., Nair, R., 2004, *Climate Policy Assessment for India: Applications of Asia–Pacific Integrated Model (AIM)*, Universities Press, New Delhi, India.

Stavins, R.N., 1998, 'What can we learn from the grand policy experiment? Lessons from SO_2 allowance trading', *Journal of Economic Perspectives* 12(3), 69–88.

Stern, N., 2006, *Stern Review: The Economics of Climate Change*, House of Lords, London.

Tata Steel, 2006, 'Indian steel outlook', in: *IISI-OECD Conference*, 16 May 2006.

TERI, 2006, *National Energy Map for India: Technology Vision 2030*, Office of the Principal Scientific Adviser, Government of India, New Delhi, India.

Turton, H., Barreto, L., 2006, 'Long-term security of energy supply and climate change', *Energy Policy* 34, 2232–2250.

UNPD (United Nations Population Division), 2006, *The World Population Prospects: The 2004 Revision Population Database*, Vol. 2006 [available at http://esa.un.org/unpp/].

WCED, 1987, *Our Common Future*, Oxford University Press, Oxford, UK.

Yergin, D., 2006, 'Ensuring energy security', *Foreign Affairs* 85(2), 69–82.

T - #0706 - 101024 - C0 - 262/190/10 - PB - 9781138002074 - Gloss Lamination